Springer Theses

Recognizing Outstanding Ph.D. Research

Aims and Scope

The series "Springer Theses" brings together a selection of the very best Ph.D. theses from around the world and across the physical sciences. Each volume has been selected for its scientific excellence by an internationally renowned institute or department. For greater accessibility to non-specialists, the published versions include an extended introduction, as well as a foreword by the student's supervisor explaining the special relevance of the work for the field. The series as a whole provides a valuable resource both for newcomers to the research fields, and for other scientists seeking detailed background information on special questions. Finally, it serves as an accredited documentation of the valuable contributions made by today's younger generation of scientists.

How to Apply

Theses prepared at internationally leading universities and institutes may be nominated for publication by supervisors with the endorsement of the head of department. Each such department is entitled to nominate one thesis per year. Enquiries and requests for a nomination form should be sent, together with the thesis, to Ms. Nathalie Berg (see link on right).

Eligibility Criteria

To qualify for consideration, theses should fulfil the following criteria:

- They must be written in good English.
- The topic should fall within the confines of Chemistry, Physics, Earth Sciences, Engineering and related interdisciplinary fields such as Materials, Nanoscience, Chemical Engineering, Complex Systems and Biophysics.
- The work reported in the thesis must represent a significant scientific advance.
- A maximum of 30% of the text may be a verbatim reproduction from the author's previous publications.
- They must have been examined and passed during the 12 months prior to nomination.
- The theses should have a clearly defined structure including an introduction accessible to new PhD students and scientists not expert in the relevant field.
- After acceptance, theses should be supplemented by a "Supervisor's Foreword" outlining the wider scientific relevance of the work.

Indexed by zbMATH.

Matthew Leighton

Stochastic Thermodynamics of Multicomponent Molecular Machines

Doctoral Thesis accepted by
Simon Fraser University, Burnaby, Canada

Springer

Author
Dr. Matthew Leighton
Department of Physics and Quantitative Biology Institute
Yale University
New Haven, CT, USA

Supervisor
Prof. David A. Sivak
Department of Physics
Simon Fraser University
Burnaby, Canada

ISSN 2190-5053 ISSN 2190-5061 (electronic)
Springer Theses
ISBN 978-3-032-02203-5 ISBN 978-3-032-02204-2 (eBook)
https://doi.org/10.1007/978-3-032-02204-2

This Springer imprint is published by the registered company Springer Nature Switzerland AG
The registered company address is: Gewerbestrasse 11, 6330 Cham, Switzerland

Supervisor's Foreword

In his Ph.D. thesis, Dr. Matthew Leighton made striking and impactful advances on several distinct yet related research fronts. His work advanced quantitative understanding of how different molecular machines interact, and how the parts of a single molecular machine coordinate their activity, to accomplish complex tasks in living organisms such as material transport and maintenance of electrical and chemical balance.

Matt derived powerful performance bounds for a remarkably broad class of collective-transport systems that are independent of model details and illuminate important performance trade-offs. He applied these bounds to estimate experimental efficiencies of molecular-machine subunits. In what I consider his most striking advance, Matt showed that all functional bipartite heat engines must operate by transmitting information between their subunits, demonstrating a fundamental linkage between heat engines and information engines; in applications he showed how widespread and vital biological machinery are unexpectedly processing large amounts of information, right under our noses. A central aspect of Matt's Ph.D. research was the development of inference tools to understand the internal thermodynamics flows in evolved molecular machines; these methods are already elucidating novel aspects of biological function at the nanoscale, and point to design principles for future engineering of synthetic nanomachines.

Matt charted his own research course with impressive independence and remarkable maturity. With only minimal steering from me, he defined his research directions, learned the background literature and mathematical techniques, investigated different theoretical directions, derived novel relations and numerically verified them, scoured the experimental literature to catalogue comparisons, and crafted six impressively engaging yet rigorous manuscripts, that were published as first-author papers in the top Physics journals and received extensive media coverage in leading popular-science outlets.

All told, this is remarkable research accomplishment—in quality, volume, and impact—for a graduate career. Matt forged a novel, multifaceted, and coherent research thrust, such that we were invited to write a review specifically on the topic of his graduate research. His research garnered awards for posters, talks, and papers, ranging from local events to international conferences and academic societies.

Outside of his thesis research, Matt was a stimulating collaborator and mentor, providing the inspiration and driving force for several other projects in my group. He was a continual font of ideas, that for years to come will help orient research in my group (and I would argue the broader research field).

Burnaby, Canada
June 2025

Prof. David A. Sivak
dsivak@sfu.ca

Parts of this thesis have been published in the following journal articles

1. **M.P. Leighton**, D.A. Sivak, "Performance Scaling and Trade-offs in Collective Motor-Driven Transport", *New Journal of Physics*, **24**:013009, 2022.
2. **M.P. Leighton**, D.A. Sivak, "Dynamic and Thermodynamic Bounds for Collective Motor-Driven Transport", *Physical Review Letters*, **129**:118102, 2022.
3. **M.P. Leighton**, D.A. Sivak, "Inferring Subsystem Efficiencies in Bipartite Molecular Machines", *Physical Review Letters*, **130**:178401, 2023.
4. **M.P. Leighton**, D.A. Sivak, "Jensen Bound for the Entropy Production Rate in Stochastic Thermodynamics", *Physical Review E (Letter)*, **109**:L012101, 2024.
5. **M.P. Leighton***, J. Ehrich*, D.A. Sivak, "Information Arbitrage in Bipartite Heat Engines", *Physical Review X*, **14**:041038, 2024.
6. **M.P. Leighton**, D.A. Sivak, "Flow of Energy and Information in Molecular Machines", *Annual Reviews of Physical Chemistry*, **76**:16.1–16.25, 2025.

Acknowledgements First and foremost, I would like to thank my supervisor David for all his support and guidance over the last 4 years. I've learned so much from you, not just about physics but also about writing concise yet impactful papers, giving comprehensible and engaging talks, and navigating the world of academia. Thanks especially for giving me total freedom to choose my own research adventure, which gave me the opportunity to develop as an independent researcher, and for encouraging and supporting me in travelling to attend scientific gatherings around the world, which allowed me to meet interesting people and significantly broaden my horizons. I could say a whole lot more here, but in short, it's been an honour working with you!

I would also like to thank my undergraduate research supervisor at Dalhousie University, Andrew Rutenberg, without whom I might never have realized that I liked doing physics research. Thanks for offering a job all those years ago to a second year student with practically zero skills and experience, and for teaching me the basics of how to go about "doing science".

Thanks to the members of my supervisory committee, Nancy Forde and John Bechhoefer, for your support and advice over the years. Your feedback and questions through committee meetings, paper proof-reading, and other discussions have helped to guide the evolution of both my research program and my career as a researcher. Thanks as well to Oren Raz, for providing valuable guidance during his time as a visiting researcher at SFU.

Special thank you to Jannik Ehrich, with whom I coathoured "the information arbitrage paper" (Chap. 7), for being a fantastic collaborator. This paper, and all the zany banter that went along with it (thanks for indulging my insistence on explaining heat engines with an analogy to sheep-trading), was hands down the most fun research project I've ever been involved in. On a more serious note, I think it's also the piece of work I'm most proud of (so far!). Thanks also to everyone else I've worked with on various smaller research projects that haven't (yet) been published, including Emma Lathouwers, Deepak Gupta, Lilian Paty, Oren Raz, Juan Carlos Perez Ramirez, Graham Rich, Johan du Buisson, John Bechhoefer, and Sara Iranbakhsh.

I would also like to thank everyone who gave feedback on the papers comprising much of the work in this thesis, including Jannik Ehrich (x4), Eric Jones (x3), Nancy Forde (x2), Shoichi Toyabe (x2), Steven Blaber (x2), John Bechhoefer (x2), Johan du Buisson (x2), Eric Werker, Oren Raz, Jordan Sawchuk, and Emanuele Penocchio.

To everyone who's been involved in the Sivak group over the last 4 years, thanks for making it a very enjoyable place to work. Thanks to Steve Large for manning the SFU booth at the CUPC grad fair in 2019, and doing a great job of advertising David's group. Thanks to Steve, Miranda, Emma, Joseph, Steven, Jannik, Eric, and Deepak, for your efforts to maintain a sense of camaraderie during my first year when everything was fully virtual due to the pandemic. Thanks to everyone who has, at some point or another, had a desk in the office for contributing to a fun working atmosphere; I've especially enjoyed the many impromptu whiteboard sessions, often not even tangentially related to research, with Miranda, Steven, Jannik, Eric, Mathis,

Hoda, Amin, Jordan, Callum, Johan, and Sara. Thanks to everyone who provided feedback on my many group meeting presentations over the years, the constant cycle of feedback and improvement has made me a much better presenter.

I thank NSERC for financial support through the CGS-M and CGS-D fellowships, as well as the government of British Columbia for support through the BC graduate scholarship. I also appreciate financial support from the SFU Physics Department, made possible through the generosity of SFU donors, including the Kirk H. Michaelian, Hargreaves, and Howard Malm graduate scholarships. Special thanks to Howard and Caroline Malm for taking a personal interest in me and my research.

Finally, on a more personal note, I would like to thank my family and friends for all their support both during my Ph.D., before, and in the next chapters of my life.

Competing Interests The author has no competing interests to declare that are relevant to the content of this manuscript.

Contents

Symbols

k_B	Boltzmann's constant
$\hbar$	Planck's constant
$x, y, \boldsymbol{x}$	Position or other degrees of freedom
t	Time
v	Velocity
T	Temperature
β	Inverse temperature $1/(k_B T)$
p	Pressure
V	Volume
N	Number of components or motors
μ	Chemical potential
$p(\cdot)$	Probability distribution
$\pi(\cdot)$	Boltzmann distribution
$S[\cdot]$	Shannon entropy
$I[X, Y]$	Mutual information
$\mathrm{TC}[\cdot]$	Total correlation
$\mathrm{Var}(\cdot)$	Variance
$\mathrm{Cov}(\cdot)$	Covariance
$\boldsymbol{m}$	Mean vector
$\boldsymbol{C}$	Covariance matrix
D	Diffusion coefficient
$\mathbf{D}$	Diffusion tensor
ζ	Friction coefficient
$\boldsymbol{\zeta}$	Friction tensor
$\boldsymbol{\mu}$	Mobility tensor
$f(\cdot)$	Force
$V(\cdot)$	Potential energy
$\xi(t)$	Gaussian white noise
$J(\cdot, t)$	Probability flux
k	Transition rates between discrete states
$\mathrm{d}e$	Infinitesimal energy change

$đq$	Infinitesimal heat
$đw$	Infinitesimal work
$\mathrm{d}s$	Infinitesimal system entropy change
$đs_r$	Infinitesimal reservoir entropy change
$đ\sigma$	Infinitesimal entropy production
$\dot{W}$	Ensemble-averaged work rate
$\dot{Q}$	Ensemble-averaged heat flow
$\dot{E}$	Ensemble-averaged energy flow
$\dot{I}$	Ensemble-averaged information flow
$\dot{S}$	Ensemble-averaged system entropy change
$\dot{\Sigma}$	Ensemble-averaged entropy production rate
$\dot{F}$	Ensemble-averaged free energy flow
$\mathcal{U}$	Energy of environment or reservoir
$\mathcal{S}$	Entropy of environment or reservoir
$\mathcal{F}$	Free energy of environment or reservoir
Φ	Massieu potential of environment or reservoir
η_{T}	Thermodynamic efficiency
η_{S}	Stokes efficiency
$\eta_{X/Y/\mathrm{M/C}}$	Subsystem efficiencies
$\langle v \rangle$	Average velocity
$v_{\max}$	Maximum velocity
D_{eff}	Effective diffusivity
D_{bare}	Bare collective diffusivity
θ	Coefficient of variation
ϕ	Fano factor
ϵ	Error
κ	Spring constant
$E^{\ddagger}$	Barrier height
$f_{\max}$	Maximum conservative force
ℓ	Period of potential energy landscape
$\boldsymbol{A}$	Matrix of linear force coefficients
g	Mobility ratio
a	Strength of non-conservative force
τ	Temperature ratio
ν	Photon frequency
n_{ν}	Photon number density
f_{ν}	Density of states
ω_{ν}	Photon angular frequency

Chapter 1
Introduction

Molecular machines transduce free energy between different forms throughout all living organisms. While truly machines in their own right, unlike their macroscopic counterparts molecular machines are characterized by stochastic fluctuations, overdamped dynamics, and soft components, and operate far from thermodynamic equilibrium. In addition, information is a relevant free-energy resource for molecular machines, leading to new modes of operation for nanoscale engines. Nonequilibrium free energy transduction in molecular machines is typically studied through the lens of stochastic thermodynamics, which permits analysis of thermodynamic quantities like work, energy, entropy, and information in nanoscale stochastic systems far from equilibrium.

Many biological and synthetic molecular machines are made up of interacting components coupled together. While individual machine components have been well studied through single-molecule experiments and computational modelling, multicomponent molecular machines are relatively underexplored. Multicomponent machines permit qualitatively new features that will be explored in this thesis, including internal flows of energy and information, and the possibility of simultaneous exposure to different sources of fluctuations.

In this thesis I apply existing and novel tools from stochastic thermodynamics to study molecular machines, with a special focus on understanding the behaviour of multicomponent molecular machines. The work in this thesis derives fundamental limits, explores model systems, and develops tools for inference from experimental data, all of which allow for novel analysis of molecular machines. Ultimately, these efforts lead to the identification of design principles which I hope will help to guide future engineering of synthetic nanomachines.

Living organisms operate far from thermodynamic equilibrium [1]. Maintaining this nonequilibrium state requires a constant input of free energy, which must in turn be converted from one form to another in order to carry out the different functions necessary for life. At the cellular level, free-energy conversion is accomplished by

M. Leighton, *Stochastic Thermodynamics of Multicomponent Molecular Machines*, Springer Theses, https://doi.org/10.1007/978-3-032-02204-2_1

molecular machines: nanoscale protein structures responsible for much of the inner workings of the cell [2].

Biological molecular machines accomplish a wide variety of tasks within cells. Most living organisms on Earth ultimately derive their energy from the Sun, with high-energy solar photons being transduced into electrochemical free energy by photosynthetic organisms using light-harvesting molecular machines such as photosystems I and II [3]. Cells typically use molecules of adenosine triphosphate (ATP) as their energy currency, which is produced primarily by ATP synthases made of coupled rotary motors that leverage cross-membrane proton gradients to synthesize ATP against its chemical-potential gradient [4]. ATP then powers a host of other molecular machines, including transport motors such as kinesin and myosin [5], protein synthesis machines such as ribosomes, trans-membrane transporters such as sodium-potassium pumps, and a wide variety of machinery for DNA replication and transcription [6]. These molecules are truly machines, taking in energy and using it to perform useful work (Fig. 1.1a).

While they are indeed machines in their own right, molecular machines inhabit a very different physical world from the macroscopic machines we interact with in our

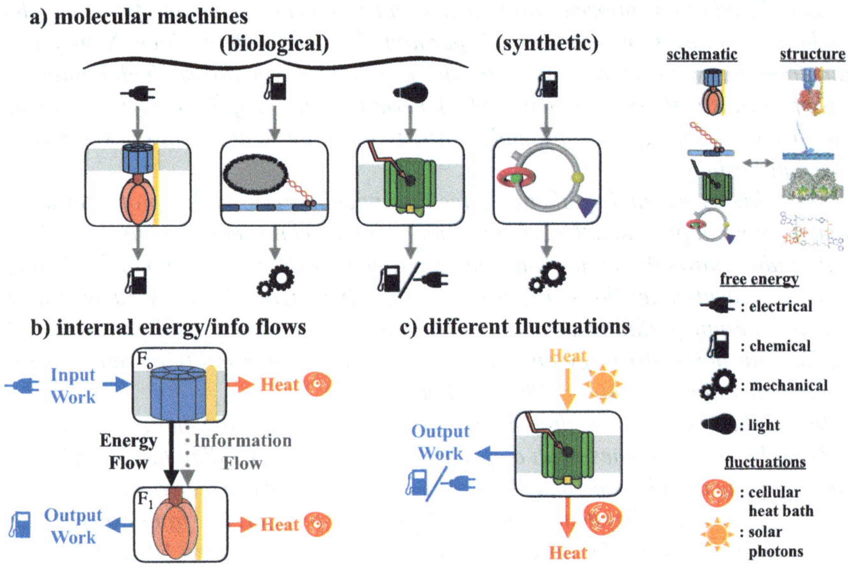

Fig. 1.1 Molecular machines transduce free energy between different forms at the nanoscale. **a** Examples of biological and synthetic molecular machines and the types of free energy they transduce: from left to right, ATP synthase, kinesin, photosystem II (all structures adapted from the Protein Data Bank [18]), and a minimal rotary motor (adapted from Ref. [19] (CC BY 4.0)). **b** Modelling ATP synthase as bipartite, made up of the coupled rotary motors F_o and F_1, allows resolution of internal flows of energy and information. **c** Some molecular machines, such as photosystem II, are in contact with multiple sources of fluctuations, allowing them to effectively act as heat engines

everyday lives [2]. First of all, their dynamics are highly overdamped, with frictional drag dominating over inertia [7]. Second, the energy scales of driving forces and interactions are comparable to the thermal energy scale $k_B T$; thus their dynamics are strongly influenced by stochastic fluctuations from the environment. Third, molecular machines are made up of many soft protein components, coupled by loose, floppy connections. Finally, their strength of driving and rapidity of operation are such that they are at all times far from thermodynamic equilibrium.

Despite these bewildering differences, molecular machines perform remarkably well – comparable to, and in some cases much better than, their macroscopic counterparts [8]. For example, photosystem II, one of the molecular machines responsible for photosynthesis in plant cells, converts solar energy into electrochemical free energy at an efficiency exceeding 40% [3]. ATP synthase, a molecular machine inside mitochondria, converts the electrical energy provided by a cross-membrane proton gradient into chemical free energy in the form of ATP, with an efficiency of 70-90% [9]. The F_o and F_1 subunits of ATP synthase are rotary motors that reach angular speeds of 150 rps [10]. Kinesin, a well-studied molecular transport motor, achieves velocities of up to 1μm (> 100 body lengths) per second [11].

For nanoscale machines where energy scales are comparable to $k_B T$, information (roughly speaking, correlated fluctuations) becomes a relevant thermodynamic resource. Molecular machines can interconvert between information and other types of free energy [12]. Conceptual models have been proposed for physically realizable molecular machines that could convert information directly into work [13], and recent experiments have realized physical information-driven engines in laboratory conditions [14–17]. Such artificial *information engines* can achieve velocities and power outputs comparable to biological molecular machines [17]. While this information-transduction mechanism is understood and demonstrated in both theory and experiment, the question remains: have biological molecular machines evolved to use information as a thermodynamic resource?

In recent years, it has become possible to design and engineer synthetic molecular machines *de novo* [20–24]; however, to date, these human-designed machines perform significantly worse than evolved biological molecular machines, for example achieving efficiencies of free-energy transduction on the order of 10^{-8} [25]. The practical promise of design principles to improve engineering of synthetic nanomachines drives the quest to better understand the inner workings of biological molecular machines. While the scientific community has learned much about biological molecular machines over the past decades, many questions remain about the physics underlying their function.

1.1 Paradigmatic Molecular Machines

Biology features a vast cornucopia of molecular machines. The molecular machines that have been explored most thoroughly through the lens of nonequilibrium thermodynamics fall into several paradigmatic classes.

1.1.1 Transport Motors

Some of the best-studied molecular motors are *transport motors* (Fig. 1.1a, 2nd panel), a large family of proteins that includes kinesins, dyneins, and myosins. Transport motors (also referred to as 'motor proteins' or just 'motors') are integral components of eukaryotic cells, with a wide range of functions including transport of large macromolecular cargo such as vesicles [26], organelles [27], or actin filaments [28], over significant distances [27]. These motors transduce chemical energy, generally in the form of ATP, into net mechanical motion in a preferred direction [2]. Indeed, motor proteins can be thought of and modelled as nanoscale thermodynamic engines, whose behaviour is characterized by stochastic mechanical and chemical dynamics [29]. Individual transport motors can reach speeds as high as $\sim 8\,\mu\text{m/s}$ [30] and make forward progress while pulling against forces on the order of $\sim 6\,\text{pN}$ [31]. Particularly well-characterized examples include kinesin and dynein motors pulling vesicles along microtubules [26], and myosin motors pulling on actin filaments to contract muscle tissue [28].

Motor proteins within cells often work collectively to transport large organelles such as mitochondria [27, 32], or even chromosomes during mitosis and meiosis [33]. Experimental determination of the number N of motors attached to a given cargo is generally challenging; nonetheless, recent studies have successfully measured N by identifying discrete peaks in a distinctly multimodal velocity distribution [34, 35] for small numbers of motors, or using more complex techniques such as quantitative immunoblots and immunoelectron microscopy [32]. Experimental investigations both *in vivo* and *in vitro* have found widely varying numbers N of motors coupled to a single cargo. In some cases, only a single motor [36] or a few motors [34] per cargo is observed, but experiments have observed as many as 200 motors bound to large organelles [32]. Likewise, in actomyosin filaments in muscle tissue, on the order of 100 motors are attached to each actin filament [37]. Collective-transport systems can also be engineered *in vitro* [38]; in this setting the number of motors can be controlled more precisely, for example using DNA scaffolds [39, 40].

1.1.2 Rotary Motors

Another critically important class of molecular machines is *rotary motors*, which feature rotational degrees of freedom driven by chemical, electrical, or mechanical forces. The family of rotary motors includes F_oF_1 ATP synthase (Fig. 1.1a, 1st panel), responsible for the synthesis of ATP within mitochondria [4]; V-ATPases, which serve as proton pumps across vacuolar membranes [41]; and the bacterial flagellar motor [42], the machinery underlying locomotion for many motile species of bacteria.

F_oF_1–ATP synthase (or just ATP synthase), an important example system in this thesis, is made up of two coupled rotary motors, F_o and F_1, both of which reversibly transduce electrochemical and mechanical free energy. ATP synthase converts electrochemical energy from a transmembrane proton gradient into the synthesis of ATP molecules via free-energy transduction between the F_o and F_1 subsystems [4]. Closely related to F_oF_1-ATP synthase, V_oV_1-ATPases are similarly composed of two coupled rotary motors and have recently been the subject of experimental investigation [43], but have yet to be studied through the lens of stochastic thermodynamics. The bacterial flagellar motor is likewise made up of multiple coupled rotary motors: a central rotor, surrounded by a varying number of stators.

1.1.3 Other Molecular Machines

Most biological free energy ultimately comes from sunlight, which is harvested by *light-harvesting molecular machines* such as photosystems I and II (Fig. 1.1a, 3rd panel) in plant cells and bacteriorhodopsin in many archaea [6]. These molecular machines take in light energy in the form of solar photons, and transduce it into various different forms of electrochemical free energy.

Biology also features a wide range of other machine types, including transmembrane pumps, transporters, and the machinery involved in DNA and RNA transcription, translation, and repair [6].

Beyond biological molecular machines, chemists and engineers have created a suite of synthetic molecular machines for various purposes [44]. These are typically inspired by biological counterparts, but to date remain far simpler and much less functional. Work in this area has yielded synthetic versions of rotary motors [21, 23] and light-driven machines [23, 45], in addition to pumps, burnt-bridge ratchets [24, 46], and assemblers. Other lines of research involve computer simulations of models for synthetic machines, allowing rapid search of design space [25, 47]. Parallel efforts to improve our understanding of biological molecular machines strive to uncover design principles to guide these engineering efforts.

1.1.4 Key Observation: Multicomponent Molecular Machines are Ubiquitous

Like many synthetic molecular machines, biological molecular machines are typically made up of interacting components coupled together. Examples include ATP synthase (composed of the coupled F_o and F_1 rotary motors [9]), the bacterial flagellar motor (a rotor coupled to $\sim$10 stators [48]), and assemblies of transport motors such as kinesin (where as many as hundreds of motors work collectively to pull large cargo [32, 37]). While single-molecule experiments and computational modelling have combined to yield an impressive understanding of how individual parts of molecular machines work in isolation [11, 49, 50], much less is known about how they interact with other components in their natural context as part of a larger conglomerate. Multicomponent molecular machines open new possibilities for free-energy transduction: in addition to exchanging free energy with external reservoirs, they also feature internal flows of free energy (comprised of energy and information) between their various coupled components [51]. Different components of multicomponent molecular machines can exchange energy with different sources of fluctuations (e.g., solar photons, active noise), leading in some cases to effective temperature differences across a machine [52] (Fig. 1.1c).

1.2 Stochastic Thermodynamics

The last 200 years have seen thermodynamics evolve from its infancy in Carnot's "Reflections on the motive power of fire" [53] to being the dominant paradigm for studying how energy moves and is transformed in systems ranging from human-created machines [54] to large-scale astronomical structures like stars [55] and even the universe itself [56], as well as small-scale molecular machines operating within living cells [2]. Historically, thermodynamics was developed to study the behaviour and performance of heat engines. The quest to design a more efficient steam engine ultimately led to the formulation and refinement of the first and second laws of thermodynamics.

As free-energy conversion devices, it is natural and appealing to view molecular machines through the lens of thermodynamics; however, they present a significant challenge for application of classical thermodynamics as a result of their far-from-equilibrium operation, few degrees of freedom, and strongly stochastic dynamics. Motivated by these difficulties, the field of *stochastic thermodynamics* has developed over the last 30 years to extend classical thermodynamics to nanoscale nonequilibrium stochastic systems. Stochastic thermodynamics provides a concrete framework for quantifying energy, heat, work, and other important quantities at the level of individual trajectories.

The first wave of important results included a collection of integrated and detailed fluctuation theorems, most notably those of Jarzynski [57] and Crooks [58]. These allowed, for the first time, direct estimation of equilibrium thermodynamic quantities like free-energy differences using data from experiments far from equilibrium [59–61]. The newfound tractability of nonequilibrium processes led to an explosion of work exploring different methods for obtaining optimal or near-optimal control protocols for probing nanoscale stochastic systems [62–64].

More recently, interest has turned to bounding and inferring entropy production of nonequilibrium systems. Results such as the thermodynamic uncertainty relations [65–67] and Jensen bounds [68, 69] provide lower bounds on entropy production rates of systems and subsystems in different contexts. In the study of molecular machines, these bounds have been used to obtain bounds and trade-offs constraining performance [65, 68, 70–72] and for thermodynamic inference of various efficiency measures [73, 74].

Central to several chapters of this thesis, the development of *bipartite* stochastic thermodynamics has enabled a finer-resolution view of energy and entropy flows into, out of, and within more complex multicomponent stochastic systems [75, 76].

This formalism has been developed primarily to study two-component molecular machines such as ATP synthase [74, 77], transport motors pulling cargo [74, 78–80], synthetic molecular motors [25], and light-driven pumps [45]. Other model systems include information engines [75, 76, 81, 82], cellular sensors [83–85], and coupled quantum dots [75]. This framework constitutes a natural lens through which to study multicomponent molecular machines, and provides a firm grounding for the study of information thermodynamics in physically realizable systems.

Studying small-scale systems also reveals the probabilistic nature of the laws of thermodynamics. This is illustrated by a famous thought experiment, known as Maxwell's demon [86]: An intelligent being with microscopic information about the position and velocity of gas molecules can separate the fast from the slow ones, apparently violating the second law. "Exorcising" Maxwell's demon [87] exposed a deep connection between information and thermodynamics [12], specifically that information about a small fluctuating system can be exploited to perform useful tasks and that this information has a fundamental cost [12, 88]. Systems that leverage information to extract work are called *information engines*, and there are abundant experimental realizations [17, 89–99].

1.3 Overview of This Thesis

In this thesis I apply existing and novel tools from stochastic thermodynamics to study molecular machines, with a special focus on understanding the behaviour of multicomponent molecular machines. At the broadest level my objective is to build a robust physical theory of biological molecular machines, guided by the following questions, which this thesis constitutes my efforts to address:

Broad Questions

1. What fundamental limits do the laws of physics impose on the performance of nanoscale machines, and how close do living organisms come to reaching these limits?
2. What can we infer about the inner workings of molecular machines from limited experimental data?
3. What design principles can we learn from biology to facilitate future engineering of synthetic nanomachines?

Throughout the discussion sections of each chapter of this thesis, boxes of the above format will highlight key results answering one or more of these questions.

This thesis is organized as follows: Chap. 2 reviews theoretical background that will be built on throughout the thesis, including bipartite (and multipartite) stochastic thermodynamics, efficiency metrics, and thermodynamic inference. Chapter 3 derives a key family of inequalities that will be used throughout the thesis, the Jensen bounds. Chaps. 4 and 5 explore the dynamics and thermodynamics of collective motor-driven transport systems, deriving scaling laws and performance trade-offs for a specific model (Chap. 4), and performance bounds and Pareto frontiers for a far more general class of models (Chap. 5). In Chap. 6 I turn to thermodynamic inference, using the Jensen bounds derived in Chap. 3 to derive upper and lower bounds on subsystem efficiencies in bipartite molecular machines. Chapters 7 and 8 consider systems in contact with multiple different sources of fluctuations, with Chap. 7 examining general bipartite heat engines while Chap. 8 focuses specifically on light-harvesting molecular machines. Finally, Chap. 9 concludes the thesis by summarizing the main results, highlighting future directions of interest, and providing a broader perspective.

1.3.1 Contributions

This Chapter along with parts of Chaps. 2 and 9 are adapted from a review article published as Ref. [100]. Chapter 3 is adapted from Ref. [69], Chap. 4 from Ref. [80], Chap. 5 from Ref. [68], Chap. 6 from Ref. [74], and Chaps. 7 and 8 from Ref. [52]. I am the sole first author for Refs. [68, 69, 74, 80, 100], in which I performed the majority of the derivations, simulations, data analysis, writing, and editing. Ref. [52] was a joint project with Jannik Ehrich, with whom I am joint first author; we both independently derived the main results, and worked together to plan and write the manuscript. The material in Chap. 7 was a full collaborative effort, while the material in Chap. 8 was developed primarily by myself in response to reviewer feedback on the initial submission of Ref. [52].

References

1. Schrödinger E (1992) What is life?: with mind and matter and autobiographical sketches. Cambridge University Press
2. Brown AI, Sivak DA (2019) Theory of nonequilibrium free energy transduction by molecular machines. Chem Rev 120(1):434–459
3. Vinyard DJ, Ananyev GM, Charles Dismukes G (2013) Photosystem II. The reaction center of oxygenic photosynthesis. Annu Rev Biochem 82:577–606
4. Oster G, Wang H (1999) ATP synthase: two motors, two fuels. Structure 7(4):R67–R72
5. Howard J (2002) Mechanics of motor proteins. In: Physics of Bio-molecules and Cells. Springer, pp 69–94
6. Phillips R, Kondev J, Theriot J, Garcia H (2012) Physical biology of the cell. Garland Science
7. Purcell EM (2014) Life at low Reynolds number. In: Physics and our world: reissue of the proceedings of a symposium in honor of Victor F Weisskopf. World Scientific, pp 47–67
8. Marden JH, Allen LR (2002) Molecules, muscles, and machines: universal performance characteristics of motors. Proc Natl Acad Sci 99(7):4161–4166
9. Silverstein TP (2014) An exploration of how the thermodynamic efficiency of bioenergetic membrane systems varies with c-subunit stoichiometry of F_0F_1 ATP synthases. J Bioenerg Biomembr 46(3):229–241
10. Akemi M-Y, Youssef H (1988) Estimation of the turnover number of bovine heart F_0F_1 complexes for ATP synthesis. Biochemistry 27(1):335–340

11. Schnitzer MJ, Block SM (1997) Kinesin hydrolyses one ATP per 8-nm step. Nature 388(6640):386–390
12. Parrondo JMR, Horowitz JM, Sagawa T (2015) Thermodynamics of information. Nat Phys 11:131
13. McGrath T, Jones NS, Wolde PRT, Ouldridge TE (2017) Biochemical machines for the interconversion of mutual information and work. Phys Rev Lett 118(2):028101
14. Serreli V, Lee C-F, Kay ER, Leigh DA (2007) A molecular information ratchet. Nature 445(7127):523–527
15. Toyabe S, Sagawa T, Ueda M, Muneyuki E, Sano M (2010) Experimental demonstration of information-to-energy conversion and validation of the generalized Jarzynski equality. Nat Phys 6(12):988–992
16. Koski JV, Maisi VF, Pekola JP, Averin DV (2014) Experimental realization of a Szilard engine with a single electron. Proc Natl Acad Sci 111(38):13786–13789
17. Saha TK, Lucero JNE, Ehrich J, Sivak DA, Bechhoefer J (2021) Maximizing power and velocity of an information engine. Proc Natl Acad Sci 118(20):e2023356118
18. Berman HM, Westbrook J, Feng Z, Gilliland G, Bhat TN, Weissig H, Shindyalov IN, Bourne PE (2000) The protein data bank. Nucl Acids Res 28(1):235–242
19. Amano S, Esposito M, Kreidt E, Leigh DA, Penocchio E, Roberts BMW (2021) Insights from an information thermodynamics analysis of a synthetic molecular motor. ChemRxiv
20. Balzani V, Credi A, Raymo FM, Fraser Stoddart J (2000) Artificial molecular machines. Angew Chem Int Ed 39(19):3348–3391
21. Wilson MR, Solà J, Carlone A, Goldup SM, Lebrasseur N, Leigh DA (2016) An autonomous chemically fuelled small-molecule motor. Nature 534(7606):235–240
22. Courbet A, Hansen J, Hsia Y, Bethel N, Park Y-J, Xu C, Moyer A, Boyken SE, Ueda G, Nattermann U et al (2022) Computational design of mechanically coupled axle-rotor protein assemblies. Science 376(6591):383–390
23. van Beek CLF, Feringa BL (2024) Coupled rotary motion in molecular motors. J Am Chem Soc
24. Korosec CS, Unksov IN, Surendiran P, Lyttleton R, Curmi PMG, Angstmann CN, Eichhorn R, Linke H, Forde NR (2024) Motility of an autonomous protein-based artificial motor that operates via a burnt-bridge principle. Nat Commun 15(1):1511
25. Amano S, Esposito M, Kreidt E, Leigh DA, Penocchio E, Roberts BMW (2022) Insights from an information thermodynamics analysis of a synthetic molecular motor. Nat Chem 14(5):530–537
26. Encalada SE, Szpankowski L, Xia C, Goldstein LSB (2011) Stable kinesin and dynein assemblies drive the axonal transport of mammalian prion protein vesicles. Cell 144(4):551–565
27. Hancock WO (2008) Intracellular transport: kinesins working together. Curr Biol 18(16):R715–R717
28. Cooke R (1997) Actomyosin interaction in striated muscle. Physiol Rev 77(3):671–697
29. Seifert U (2012) Stochastic thermodynamics, fluctuation theorems and molecular machines. Rep Prog Phys 75(12):126001
30. Purcell TJ, Naber N, Franks-Skiba K, Dunn AR, Eldred CC, Berger CL, Málnási-Csizmadia A, Spudich JA, Swank DM, Pate E et al (2011) Nucleotide pocket thermodynamics measured by EPR reveal how energy partitioning relates myosin speed to efficiency. J Mol Biol 407(1):79–91
31. Fallesen TL, Macosko JC, Holzwarth G (2011) Force-velocity relationship for multiple kinesin motors pulling a magnetic bead. Eur Biophys J 40(9):1071–1079
32. Leopold PL, McDowall AW, Kevin Pfister K, Bloom GS, Brady ST (1992) Association of kinesin with characterized membrane-bounded organelles. Cell Motil Cytoskelet 23(1):19–33
33. Hatsumi M, Endow SA (1992) Mutants of the microtubule motor protein, nonclaret disjunctional, affect spindle structure and chromosome movement in meiosis and mitosis. J Cell Sci 101(3):547–559
34. Shtridelman Y, Cahyuti T, Townsend B, DeWitt D, Macosko JC (2008) Force-velocity curves of motor proteins cooperating in vivo. Cell Biochem Biophys 52(1):19–29

35. Shtridelman Y, Holzwarth GM, Bauer CT, Gassman NR, DeWitt DA, Macosko JC (2009) In vivo multimotor force-velocity curves by tracking and sizing sub-diffraction limited vesicles. Cell Mol Bioeng 2(2):190–199
36. Svoboda K, Block SM (1994) Force and velocity measured for single kinesin molecules. Cell 77(5):773–784
37. Rastogi K, Puliyakodan MS, Pandey V, Nath S, Elangovan R (2016) Maximum limit to the number of myosin II motors participating in processive sliding of actin. Sci Rep 6(1):1–10
38. DelRosso NV, Derr ND (2017) Exploiting molecular motors as nanomachines: the mechanisms of de novo and re-engineered cytoskeletal motors. Curr Opin Biotechnol 46:20–26
39. Furuta K, Furuta A, Toyoshima YY, Amino M, Oiwa K, Kojima H (2013) Measuring collective transport by defined numbers of processive and nonprocessive kinesin motors. Proc Natl Acad Sci 110(2):501–506
40. Derr ND, Goodman BS, Jungmann R, Leschziner AE, Shih WM, Reck-Peterson SL (2012) Tug-of-war in motor protein ensembles revealed with a programmable DNA origami scaffold. Science 338(6107):662–665
41. Beyenbach KW, Wieczorek H (2006) The V-type H+ ATPase: molecular structure and function, physiological roles and regulation. J Exp Biol 209(4):577–589
42. Sowa Y, Berry RM (2008) Bacterial flagellar motor. Q Rev Biophys 41(2):103–132
43. Kosugi T, Iida T, Tanabe M, Iino R, Koga N (2023) Design of allosteric sites into rotary motor V_1-ATPase by restoring lost function of pseudo-active sites. Nat Chem 15(11):1591–1598
44. Erbas-Cakmak S, Leigh DA, McTernan CT, Nussbaumer AL (2015) Artificial molecular machines. Chem Rev 115(18):10081–10206
45. Corra S, Bakić MT, Groppi J, Baroncini M, Silvi S, Penocchio E, Esposito M, Credi A (2022) Kinetic and energetic insights into the dissipative non-equilibrium operation of an autonomous light-powered supramolecular pump. Nat Nanotechnol 17(7):746–751
46. Bazrafshan A, Kyriazi M-E, Holt BA, Deng W, Piranej S, Hanquan S, Yuesong H, El-Sagheer AH, Brown T, Kwong GA et al (2021) DNA gold nanoparticle motors demonstrate processive motion with bursts of speed up to 50 nm per second. ACS Nano 15(5):8427–8438
47. Albaugh A, Gingrich TR (2022) Simulating a chemically fueled molecular motor with nonequilibrium molecular dynamics. Nat Commun 13(1):2204
48. Meacci G, Yuhai T (2009) Dynamics of the bacterial flagellar motor with multiple stators. Proc Natl Acad Sci 106(10):3746–3751
49. Visscher K, Schnitzer MJ, Block SM (1999) Single kinesin molecules studied with a molecular force clamp. Nature 400(6740):184–189
50. Toyabe S, Okamoto T, Watanabe-Nakayama T, Taketani H, Kudo S, Muneyuki E (2010) Nonequilibrium energetics of a single F_1-ATPase molecule. Phys Rev Lett 104(19):198103
51. Ehrich J, Sivak D (2023) Energy and information flows in autonomous systems. Front Phys 11:1108357
52. Leighton MP, Ehrich J, Sivak DA (2024) Information arbitrage in bipartite heat engines. Phys Rev X 14:041038
53. Carnot S (1824) Reflections on the motive power of fire, and on machines fitted to develop that power. Paris Bachelier 108:1824
54. Rajput RK (2009) Engineering thermodynamics: a computer approach (SI units version). Jones & Bartlett Publishers
55. LeBlanc F (2011) An introduction to stellar astrophysics. John Wiley & Sons
56. Weinberg S (2008) Cosmology. OUP Oxford
57. Jarzynski C (1997) Nonequilibrium equality for free energy differences. Phys Rev Lett 78(14):2690
58. Crooks GE (1999) Entropy production fluctuation theorem and the nonequilibrium work relation for free energy differences. Phys Rev E 60(3):2721
59. Hummer G, Szabo A (2001) Free energy reconstruction from nonequilibrium single-molecule pulling experiments. Proc Natl Acad Sci 98(7):3658–3661
60. Liphardt J, Dumont S, Smith SB, Jr IT, Bustamante C (2002) Equilibrium information from nonequilibrium measurements in an experimental test of Jarzynski's equality. Science 296(5574):1832–1835

61. Collin D, Ritort F, Jarzynski C, Smith SB, Jr IT, Bustamante C (2005) Verification of the Crooks fluctuation theorem and recovery of RNA folding free energies. Nature 437(7056):231–234
62. Schmiedl T, Seifert U (2007) Optimal finite-time processes in stochastic thermodynamics. Phys Rev Lett 98(10):108301
63. Sivak DA, Crooks GE (2012) Thermodynamic metrics and optimal paths. Phys Rev Lett 108(19):190602
64. Aurell E, Mejía-Monasterio C, Muratore-Ginanneschi P (2011) Optimal protocols and optimal transport in stochastic thermodynamics. Phys Rev Lett 106(25):250601
65. Barato AC, Seifert U (2015) Thermodynamic uncertainty relation for biomolecular processes. Phys Rev Lett 114(15):158101
66. Gingrich TR, Horowitz JM, Perunov N, England JL (2016) Dissipation bounds all steady-state current fluctuations. Phys Rev Lett 116(12):120601
67. Horowitz JM, Gingrich TR (2020) Thermodynamic uncertainty relations constrain non-equilibrium fluctuations. Nat Phys 16(1):15–20
68. Leighton MP, Sivak DA (2022) Dynamic and thermodynamic bounds for collective motor-driven transport. Phys Rev Lett 129(11):118102
69. Leighton MP, Sivak DA (2024) Jensen bound for the entropy production rate in stochastic thermodynamics. Phys Rev E 109(1):L012101
70. Seifert U (2018) Stochastic thermodynamics: from principles to the cost of precision. Phys Stat Mech Appl 504:176–191
71. Song Y, Hyeon C (2020) Thermodynamic cost, speed, fluctuations, and error reduction of biological copy machines. J Phys Chem Lett 11(8):3136–3143
72. Albaugh A, Rueih-Sheng F, Geyao G, Gingrich TR (2023) Limits on the precision of catenane molecular motors: insights from thermodynamics and molecular dynamics simulations. J Chem Theory Comput 20(1):1–6
73. Pietzonka P, Barato AC, Seifert U (2016) Universal bound on the efficiency of molecular motors. J Stat Mech Theory Exp 12:124004
74. Leighton MP, Sivak DA (2023) Inferring subsystem efficiencies in bipartite molecular machines. Phys Rev Lett 130(17):178401
75. Horowitz JM, Esposito M (2014) Thermodynamics with continuous information flow. Phys Rev X 4(3):031015
76. Barato AC, Seifert U (2017) Thermodynamic cost of external control. New J Phys 19(7):073021
77. Lathouwers E, Lucero JNE, Sivak DA (2020) Nonequilibrium energy transduction in stochastic strongly coupled rotary motors. J Phys Chem Lett 11(13):5273–5278
78. Zimmermann E, Seifert U (2015) Effective rates from thermodynamically consistent coarse-graining of models for molecular motors with probe particles. Phys Rev E 91(2):022709
79. Brown AI, Sivak DA (2019) Pulling cargo increases the precision of molecular motor progress. EPL (Eur Lett) 126(4):40004
80. Leighton MP, Sivak DA (2022) Performance scaling and trade-offs for collective motor-driven transport. New J Phys 24(1):013009
81. Horowitz JM, Sandberg H (2014) Second-law-like inequalities with information and their interpretations. New J Phys 16(12):125007
82. Ehrich J, Still S, Sivak DA (2023) Energetic cost of feedback control. Phys Rev Res 5(2):023080
83. Still S, Sivak DA, Bell AJ, Crooks GE (2012) Thermodynamics of prediction. Phys Rev Lett 109:120604
84. Barato AC, Hartich D, Seifert U (2014) Efficiency of cellular information processing. New J Phys 16:103024
85. Hartich D, Barato AC, Seifert U (2016) Sensory capacity: an information theoretical measure of the performance of a sensor. Phys Rev E 93:022116
86. Knott CG (1911) Life and scientific work of peter Guthrie Tait. Cambridge University Press, London

87. Leff HS, Rex AF (eds) (2003) Maxwell's demon 2. IOP Publishing
88. Landauer R (1961) Irreversibility and heat generation in the computing process. IBM J Res Dev 5(3):183–191
89. Toyabe S, Sagawa T, Ueda M, Muneyuki E, Sano M (2010) Experimental demonstration of information-to-energy conversion and validation of the generalized Jarzynski equality. Nat Phys 6(12):988
90. Camati PA, Peterson JPS, Batalhao TB, Micadei K, Souza AM, Sarthour RS, Oliveira IS, Serra RM (2016) Experimental rectification of entropy production by Maxwell's demon in a quantum system. Phys Rev Lett 117(24):240502
91. Cottet N, Jezouin S, Bretheau L, Campagne-Ibarcq P, Ficheux Q, Anders J, Auffèves A, Azouit R, Rouchon P, Huard B (2017) Observing a quantum Maxwell demon at work. Proc Natl Acad Sci 114(29):7561
92. Masuyama Y, Funo K, Murashita Y, Noguchi A, Kono S, Tabuchi Y, Yamazaki R, Ueda M, Nakamura Y (2018) Information-to-work conversion by Maxwell's demon in a superconducting circuit quantum electrodynamical system. Nat Commun 9(1):1
93. Koski JV, Maisi VF, Pekola JP, Averin DV (2014) Experimental realization of a Szilard engine with a single electron. Proc Natl Acad Sci 111(38):13786
94. Chida K, Desai S, Nishiguchi K, Fujiwara A (2017) Power generator driven by Maxwell's demon. Nat Commun 8(1):1
95. Admon T, Rahav S, Roichman Y (2018) Experimental realization of an information machine with tunable temporal correlations. Phys Rev Lett 121(18):180601
96. Paneru G, Lee DY, Tlusty T, Pak HK (2018) Lossless Brownian information engine. Phys Rev Lett 120(2):020601
97. Paneru G, Lee DY, Park J-M, Park JT, Noh JD, Pak HK (2018) Optimal tuning of a Brownian information engine operating in a nonequilibrium steady state. Phys Rev E 98(5):052119
98. Ribezzi-Crivellari M, Ritort F (2019) Large work extraction and the Landauer limit in a continuous Maxwell demon. Nat Phys 15(7):660–664
99. Paneru G, Dutta S, Sagawa T, Tlusty T, Pak HK (2020) Efficiency fluctuations and noise induced refrigerator-to-heater transition in information engines. Nat Commun 11(1):1–8
100. Leighton MP, Sivak DA (2025) Flow of energy and information in molecular machines. Annu Rev Phys Chem 76

Chapter 2
Theoretical Background

2.1 Motivation and Overview

Classical thermodynamics is primarily concerned with heat engines operating cyclically in the quasistatic limit, or systems coming to equilibrium with external reservoirs; by contrast, molecular machines operate far from thermodynamic equilibrium. We are typically interested in their behaviour in the long-time limit, which is dominated by the steady state. Thus, modelling molecular machines within the cellular environment requires a thermodynamics of nonequilibrium steady states.

As a paradigm for mmachines, consider a nanoscale stochastic system in contact with an equilibrium thermal reservoir at temperature T (and inverse temperature $\beta \equiv 1/(k_{\mathrm{B}}T)$) with which it can exchange heat, as well as nonequilibrium free-energy reservoirs with which it can exchange work. Consider F_oF_1-ATP synthase as a specific example, in which case the nonequilibrium free-energy reservoirs are the hydrogen-ion gradient across the mitochondrial membrane and nonequilibrium concentrations of ATP and ADP.

In the sections that follow, and indeed throughout most of this thesis, I focus on systems that can be described by continuous degrees of freedom. That said, the theory of stochastic thermodynamics developed throughout this chapter can equivalently be applied to systems with discrete degrees of freedom, indeed Chaps. 7 and 8 will feature examples of such systems. For an overview of stochastic thermodynamics for discrete stochastic systems, I refer the interested reader to Ref. [1].

M. Leighton, *Stochastic Thermodynamics of Multicomponent Molecular Machines*, Springer Theses, https://doi.org/10.1007/978-3-032-02204-2_2

2.2 Random Variables, Probability, and Information Theory

Information as a thermodynamic resource is an important theme throughout this thesis. Information can be quantitatively characterized using tools from *information theory* [2]. Most centrally, information theory is concerned with quantifying uncertainty in random variables.

Throughout this thesis, I will refer to random variables with capital letters (e.g., X), and specific instances of a random variable with lower-case letters (e.g., x). The probability distribution of a random variable will be denoted $p(x)$, and ensemble averages over $p(x)$ (or the relevant probability distribution) denoted by angle brackets:

$$\langle g(x) \rangle \equiv \int g(x) p(x) \mathrm{d}x. \tag{2.1}$$

Entropy is a concept central to both statistical physics and information theory. For a single random variable, the Shannon entropy is

$$S[X] \equiv \langle -\ln p(x) \rangle , \tag{2.2}$$

while for two variables it is:

$$S[X, Y] \equiv \langle -\ln p(x, y) \rangle . \tag{2.3}$$

For two variables, the relevant distribution is the joint distribution $p(x, y)$. However, the uncertainty about individual variables can be quantified by the marginal distributions,

$$p(x) \equiv \int p(x, y) \mathrm{d}y, \tag{2.4a}$$

$$p(y) \equiv \int p(x, y) \mathrm{d}x. \tag{2.4b}$$

Additionally, the dependence of one variable on another can be quantified by the conditional distributions,

$$p(x|y) \equiv p(x, y)/p(y), \tag{2.5a}$$

$$p(y|x) \equiv p(x, y)/p(x). \tag{2.5b}$$

These distributions have their own entropies, for example the marginal entropy of X and conditional entropy of X given Y are

$$S[X] \equiv \langle -\ln p(x) \rangle , \tag{2.6a}$$

$$S[X|Y] \equiv \langle -\ln p(x|y) \rangle , \tag{2.6b}$$

and similarly for Y. Here, and elsewhere where there are two random variables under consideration, angle brackets denote averages over the distribution $p(x, y)$.

For two variables, another important information-theoretic quantity is the mutual information, which is the difference between the marginal and conditional entropies:

$$\begin{aligned} I[X, Y] &\equiv S[X] - S[X|Y] \\ &= S[Y] - S[Y|X] \\ &= \left\langle \ln \frac{p(x, y)}{p(x)p(y)} \right\rangle \end{aligned} \tag{2.7}$$

The mutual information is a measure of how much the uncertainty about one variable is reduced by knowledge of the other variable.

The Shannon entropy extends trivially to N variables; for $\boldsymbol{X} = \{X_1, ..., X_N\}$, the Shannon entropy is

$$S[\boldsymbol{X}] \equiv \langle -\ln p(\boldsymbol{x}) \rangle . \tag{2.8}$$

There is no unique generalization of the mutual information beyond two random variables. Ref. [3] reviews several of the most commonly used possibilities. For the purpose of this thesis, the only multivariate information measure considered will be the total correlation,

$$\mathrm{TC}[\boldsymbol{X}] \equiv \sum_{i=1}^{N} S[X_i] - S[\boldsymbol{X}]. \tag{2.9}$$

2.3 Stochastic Thermodynamics for One Degree of Freedom

Consider a stochastic system coupled to a thermal reservoir at temperature T, with continuous state quantified by the random variable X. The particle experiences a potential energy landscape $V(x)$ and nonconservative force $f_{\text{nc}}(x)$. In the overdamped limit, the dynamics of the particle are described by the overdamped Langevin equation:

$$\dot{x} = \frac{1}{\zeta} \underbrace{[f_{\text{nc}}(x) - \partial_x V(x)]}_{f(x)} + \sqrt{2D}\xi(t). \tag{2.10}$$

Here ζ and D are the friction and diffusion coefficients respectively, and $f(x) = f_{\text{nc}}(x) - \partial_x V(x)$ is the total force. Finally, $\xi(t)$ denotes Gaussian white noise, which is defined by the statistical properties

$$\langle \xi(t) \rangle = 0, \tag{2.11a}$$

$$\left\langle \xi(t)\xi(t') \right\rangle = \delta(t - t'). \tag{2.11b}$$

The Langevin equation describes individual trajectories, which are stochastic. By contrast, the probability distribution $p(x, t)$ evolves deterministically according to the Fokker-Planck equation:

$$\partial_t p(x, t) = -\partial_x J(x, t), \tag{2.12}$$

for probability flux

$$J(x, t) = \frac{1}{\zeta} f(x) p(x, t) - D\partial_x p(x, t). \tag{2.13}$$

Most informatively, Langevin dynamics can be described at the level of a probability distribution over the space of all possible trajectories. This can be done using a stochastic path-integral formalism [4] which, while mathematically fascinating, will not be used in this thesis.

As a paradigmatic example, consider a colloidal particle on a ring, with position quantified by $x \in [0, 2\pi]$. In the absence of a nonconservative force ($f_{\mathrm{nc}} = 0$), the particle should relax to the equilibrium Boltzmann distribution, which we denote $\pi(x)$,

$$\lim_{t\to\infty} p(x, t) = \pi(x) \propto \exp\left(\frac{-V(x)}{k_{\mathrm{B}}T}\right). \tag{2.14}$$

For this to be the case requires

$$D\zeta = k_{\mathrm{B}}T. \tag{2.15}$$

Equation (2.15) is an example of a fluctuation-dissipation theorem (FDT) and was derived by Einstein [5].

If instead $f_{\mathrm{nc}} \neq 0$, then the system will not reach equilibrium. For a colloidal particle on a ring, $p(x, t)$ will converge to a nonequilibrium steady state (NESS), where

$$\lim_{t\to\infty} p(x, t) = p(x) \neq \pi(x). \tag{2.16}$$

The goal of stochastic thermodynamics is to build a quantitative thermodynamic description of microscopic stochastic systems, including first and second laws governing flows of energy and entropy. To do this, we must first define these quantities.

2.3.1 Energy Flows and the First Law

In the absence of kinetic energy (since the dynamics are overdamped, momentum does not persist over significant timescales), the total energy of the particle in microstate x is simply given by its potential energy,

$$e(x) = V(x). \tag{2.17}$$

Suppose the particle moves by an infinitesimally small step $\mathrm{d}x$. Then the energy changes by

$$\mathrm{d}e = \partial_x V(x) \circ \mathrm{d}x. \tag{2.18}$$

Here '$\circ$' denotes multiplication in the Stratonovich sence, which is explained in the box below.

Stratonovich Discretization and Ensemble Averages When multiplying x-dependent quantities by changes in x (either $\mathrm{d}x$ or $\dot{x}$), I use the symbol $\circ$ to indicate that the multiplication should be interpreted in the Stratonovich sense. This means that products like $f(x) \circ \mathrm{d}x$ are discretized in the form

$$f(x) \circ \mathrm{d}x = f(x) \cdot [x(t + \mathrm{d}t/2) - x(t - \mathrm{d}t/2)] \,. \tag{2.19}$$

Using the Stratonovich convention, rather than its alternative the Itô convention, ensures that the chain rule of calculus holds as in deterministic calculus. For further discussion of this technical detail, I refer the interested reader to Ref. [6].

The Stratonovich discretization also makes possible the evaluation of ensemble averages of the form

$$\dot{G} \equiv \langle g(x) \circ \dot{x} \rangle_{p(\dot{x},x)} \,. \tag{2.20}$$

As proven, for example by Ehrich [7], these can be evaluated as

$$\langle g(x) \circ \dot{x} \rangle_{p(\dot{x},x)} = \left\langle g(x) \frac{J(x)}{p(x)} \right\rangle_{p(x)} \,. \tag{2.21}$$

The work done on the particle by the nonconservative force can also be quantified as

$$đw = f_{\mathrm{nc}}(x) \circ \mathrm{d}x. \tag{2.22}$$

Over the course of the small step $\mathrm{d}x$, energy must be conserved, so the balance of the infinitesimal energy change and work must equal the heat exchanged by the particle with its environment. This leads to the natural definition of the heat,

$$đq = -f(x) \circ \mathrm{d}x. \tag{2.23}$$

With this definition, we recover a first law describing energy balance for an infinitesimal state change $\mathrm{d}x$,

$$\mathrm{d}e = đw + đq. \tag{2.24}$$

Integrating this equation over a trajectory, $x(\cdot) = \{x(t), 0 \leq t \leq \tau\}$, yields a first law for trajectories

$$e(\tau) - e(0) = \int_0^\tau de[x(t)] \tag{2.25a}$$

$$= \int_0^\tau đw[x(t)] + \int_0^\tau đq[x(t)] \tag{2.25b}$$

$$= w[x(\cdot)] + q[x(\cdot)]. \tag{2.25c}$$

2.3.2 Entropy Flows and the Second Law

With the first law now established, it is time to turn to the second law. First, we must define entropy. To do this, first consider the entropy of an ensemble of systems, given by the Shannon entropy (2.2). For a single member of the ensemble, then, defining the state-dependent entropy as [8]

$$s(x) = -\ln p(x), \tag{2.26}$$

yields the Shannon entropy when the ensemble average is taken,

$$S[X] = \langle s(x) \rangle . \tag{2.27}$$

Again considering an infinitesimal change in the system state $\mathrm{d}x$, the infinitesimal change in the entropy of the universe (here and throughout, we measure entropy production in dimensionless units scaled by Boltzmann's constant k_B per second) is [8]

$$đ\sigma = đs_r + \mathrm{d}s, \tag{2.28}$$

where s_r is the entropy of the reservoir. This follows naturally from considering the entire universe to comprise nothing but the system and reservoir. Since the reservoir is assumed to be at equilibrium, the change in its entropy is the heat divided by the temperature,

$$đs_r = -đq/(k_\mathrm{B}T). \tag{2.29}$$

The change in the system entropy is more subtle, but can be derived as follows [8]:

$$\mathrm{d}s = -\partial_x \ln p(x) \tag{2.30a}$$

$$= -\frac{\partial_x p(x)}{p(x)} \circ \mathrm{d}x \tag{2.30b}$$

$$= \frac{J(x)}{Dp(x)} \circ \mathrm{d}x - \frac{f(x)}{k_\mathrm{B}T} \circ \mathrm{d}x \tag{2.30c}$$

$$= \frac{J(x)}{Dp(x)} \circ \mathrm{d}x + \text{đ}q / (k_B T) \,. \tag{2.30d}$$

Here in the third line I inserted the Fokker-Planck equation, and in the fourth line I identified the heat from its definition. Combining the results for $\text{đ}s_r$ and $\mathrm{d}s$, I obtain the change in the entropy of the universe:

$$\text{đ}\sigma = \frac{J(x)}{Dp(x)} \circ \mathrm{d}x \,. \tag{2.31}$$

Note that $\text{đ}\sigma$ need not be non-negative! This means that individual trajectories can "break the second law", decreasing the entropy of the universe. The next subsection shows how the second law is recovered from this quagmire.

2.3.3 Ensemble Averages

For molecular machines, as with many other stochastic systems of interest, we are generally interested in ensemble-averaged rates of energetic and entropic quantities at steady state.

For example, using Eq. (2.21), the ensemble-averaged work rate is

$$\dot{W} = \left\langle \frac{\text{đ}w}{\mathrm{d}t} \right\rangle \tag{2.32a}$$

$$= \langle f_{\mathrm{nc}}(x) \circ \dot{x} \rangle_{p(x,\dot{x})} \tag{2.32b}$$

$$= \left\langle f_{\mathrm{nc}}(x) \frac{J(x)}{p(x)} \right\rangle . \tag{2.32c}$$

Similarly, the heat and energy flows are

$$\dot{Q} = -\left\langle f(x) \frac{J(x)}{p(x)} \right\rangle , \tag{2.33a}$$

$$\dot{E} = \left\langle \partial_x V(x) \frac{J(x)}{p(x)} \right\rangle . \tag{2.33b}$$

Thus, as expected, the first law extends from trajectory-level to ensemble-averaged energy flows:

$$\dot{E} = \dot{Q} + \dot{W} \,. \tag{2.34}$$

At steady state, the energy is constant, so that $\dot{Q} + \dot{W} = 0$.

Performing the same ensemble averages yields average rates of change for the system entropy (S) and entropy of the universe (Σ):

$$\dot{S} = -\left\langle \partial_x \ln p(x) \frac{J(x)}{p(x)} \right\rangle, \tag{2.35a}$$

$$\dot{\Sigma} = \left\langle \frac{J(x)^2}{Dp(x)^2} \right\rangle. \tag{2.35b}$$

Thus the second law at the ensemble level is

$$\dot{\Sigma} = \dot{S} - \dot{Q}/(k_B T) \geq 0. \tag{2.36}$$

Here nonnegativity of the entropy production follows from the square inside the ensemble average.

2.4 Two Degrees of Freedom: Bipartite Systems

As a next step, consider now a system which can be decomposed into two *subsystems*, with two stochastic degrees of freedom evolving according to coupled overdamped Langevin equations:

$$\dot{x} = \frac{1}{\zeta_X} \underbrace{\left[f_X^{\text{nc}}(x, y) - \partial_x V(x, y) \right]}_{f_X(x,y)} + \sqrt{2D_X} \xi_X(t), \tag{2.37a}$$

$$\dot{y} = \frac{1}{\zeta_Y} \underbrace{\left[f_Y^{\text{nc}}(x, y) - \partial_y V(x, y) \right]}_{f_Y(x,y)} + \sqrt{2D_Y} \xi_Y(t). \tag{2.37b}$$

As in the one-dimensional case, each of the noise terms $\xi_X(t)$ and $\xi_Y(t)$ satisfy Eq. (2.11). In addition, in the sections that follow (and throughout this thesis) I assume that the dynamics of X and Y are *bipartite* [9–11], meaning that each subsystem is in contact with different thermal reservoirs which may in general be at different temperatures T_X and T_Y. For Langevin dynamics, this requires the two noise terms to be uncorrelated,

$$\langle \xi_X(t) \xi_Y(t') \rangle = 0, \quad \forall t, t'. \tag{2.38}$$

Again, I assume the fluctuation-dissipation theorem holds for each subsystem, so that

$$\zeta_X D_X = k_B T_X, \tag{2.39a}$$

$$\zeta_Y D_Y = k_B T_Y. \tag{2.39b}$$

At the distribution level, these coupled Langevin dynamics are described by a single Fokker-Planck equation,

$$\partial_t p(x, y, t) = -\partial_x J_X(x, y, t) - \partial_y J_Y(x, y, t), \tag{2.40}$$

for probability fluxes

$$J_X(x, y, t) = \frac{1}{\zeta_X}\left[f_X^{\mathrm{nc}}(x, y) - \partial_x V(x, y)\right] p(x, y, t) - D_X \partial_x p(x, y, t), \tag{2.41a}$$

$$J_Y(x, y, t) = \frac{1}{\zeta_Y}\left[f_Y^{\mathrm{nc}}(x, y) - \partial_y V(x, y)\right] p(x, y, t) - D_Y \partial_y p(x, y, t). \tag{2.41b}$$

For Fokker-Planck dynamics, the bipartite assumption requires that there are no terms of the form $\partial_x \partial_y p$.

2.4.1 Energy Flows and First Laws

We once again consider the change in system energy $e(x, y) = V(x, y)$ over an infinitesimal time interval $\mathrm{d}t$, this time with both X and Y changing by respective amounts $\mathrm{d}x$ and $\mathrm{d}y$. The change in energy is

$$\mathrm{d}e = \underbrace{\partial_x V \circ \mathrm{d}x}_{\mathrm{d}e_X} + \underbrace{\partial_y V \circ \mathrm{d}y}_{\mathrm{d}e_Y} = \underbrace{f_X^{\mathrm{nc}} \circ \mathrm{d}x}_{\text{đ}w_X} + \underbrace{f_Y^{\mathrm{nc}} \circ \mathrm{d}y}_{\text{đ}w_Y} \underbrace{- \left[f_X^{\mathrm{nc}} - \partial_x V\right] \circ \mathrm{d}x}_{\text{đ}q_X} \underbrace{- \left[f_Y^{\mathrm{nc}} - \partial_y V\right] \circ \mathrm{d}y}_{\text{đ}q_Y}. \tag{2.42}$$

Here, we have decomposed the total change in system energy into work and heat contributions from each of the X and Y subsystems. From Eq. (2.42), we can see that more fine-grained relationships hold between the changes in energy, works, and heats for each of X and Y:

$$\mathrm{d}e_X = \text{đ}w_X + \text{đ}q_X, \tag{2.43a}$$

$$\mathrm{d}e_Y = \text{đ}w_Y + \text{đ}q_Y. \tag{2.43b}$$

These are the *subsystem first laws*, here written at the trajectory level.

As in Sect. 2.3, considering rates and taking ensemble averages gives first laws describing average rates of energy flow for each subsystem:

$$\dot{E}_X = \dot{W}_X + \dot{Q}_X, \tag{2.44a}$$

$$\dot{E}_Y = \dot{W}_Y + \dot{Q}_Y. \tag{2.44b}$$

At steady state, the energy is constant, so that any increase in energy due to Y dynamics must be accompanied by a matching decrease in energy due to X: $\dot{E}_X =$

$-\dot{E}_Y$. Treating Y as an external control parameter driving X, we can also think about the energy flow $\dot{E}_Y$ as "transduced work" from Y to X [1].

2.4.2 Entropy Flows and Second Laws

As before, the change $\mathrm{d}\sigma$ in the entropy of the universe is the sum of the change in system energy and change in reservoir entropies,

$$\mathrm{d}\sigma = \mathrm{d}s + \mathrm{d}s_r \tag{2.45a}$$

$$= \underbrace{-\partial_x \ln p(x,y) \circ \mathrm{d}x}_{\mathrm{d}s_X} \underbrace{-\partial_y \ln p(x,y) \circ \mathrm{d}y}_{\mathrm{d}s_Y} + \underbrace{\beta_X \left[f_X^{\mathrm{nc}} - \partial_x V\right] \circ \mathrm{d}x}_{-\beta_X đq_X} + \underbrace{\beta_Y \left[f_Y^{\mathrm{nc}} - \partial_y V\right] \circ \mathrm{d}y}_{-\beta_Y đq_Y}. \tag{2.45b}$$

Here in the last line I split the entropy changes into contributions due to each of X and Y. With this splitting, the total entropy production can be decomposed into contributions from each subsystem, $\mathrm{d}\sigma = \mathrm{d}\sigma_X + \mathrm{d}\sigma_Y$, given by

$$\mathrm{d}\sigma_X = \mathrm{d}s_X - \beta_X đq_X, \tag{2.46a}$$

$$\mathrm{d}\sigma_Y = \mathrm{d}s_Y - \beta_Y đq_Y. \tag{2.46b}$$

For ensemble-averaged rates, this gives

$$\dot{\Sigma}_Y = \dot{S}_X[X,Y] - \beta_Y \dot{Q}_Y \geq 0, \tag{2.47a}$$

$$\dot{\Sigma}_X = \dot{S}_Y[X,Y] - \beta_X \dot{Q}_X \geq 0. \tag{2.47b}$$

Here we now find that the subsystem entropy production rates $\dot{\Sigma}_X$ and $\dot{\Sigma}_Y$ are both nonnegative.

2.4.3 Information Flow

Using the definition of the mutual information (2.7), the entropy rates $\dot{S}_X$ and $\dot{S}_Y$ can be decomposed further:

$$\dot{S}_X[X,Y] = \langle -\partial_x \ln p(x,y) \circ \dot{x} \rangle \tag{2.48a}$$

$$= \left\langle -\partial_x \ln \left[p(x) \frac{p(x,y)}{p(x)} \right] \circ \dot{x} \right\rangle \tag{2.48b}$$

$$= \langle -\partial_x \ln p(x) \circ \dot{x} \rangle - \left\langle \partial_x \ln \frac{p(x,y)}{p(x)} \circ \dot{x} \right\rangle \tag{2.48c}$$

$$= \langle -\partial_x \ln p(x) \circ \dot{x} \rangle - \left\langle \partial_x \ln \frac{p(x,y)}{p(x)p(y)} \circ \dot{x} \right\rangle \tag{2.48d}$$

$$= \mathrm{d}_t S[X] - \dot{I}_X . \tag{2.48e}$$

Here $\mathrm{d}_t S[X]$ is the rate of change of the marginal entropy $S[X]$, and $\dot{I}_X$ is the rate at which the dynamics of X change the mutual information, which we call the *information flow* (alternatively known as the learning rate [12]). $\dot{S}_Y$ can be decomposed similarly. At steady state, the marginal entropies $S[X]$ and $S[Y]$ are constant, so that $\dot{S}_X = -\dot{I}_X$ and $\dot{S}_Y = -\dot{I}_Y$. Thus, the second laws can be rewritten in terms of the information flow as [11]

$$\dot{\Sigma}_Y = -\beta_Y \dot{Q}_Y - \dot{I}_Y \geq 0, \tag{2.49a}$$

$$\dot{\Sigma}_X = -\beta_X \dot{Q}_X - \dot{I}_X \geq 0. \tag{2.49b}$$

As with the energy flow, at steady state the mutual information is constant, so that $\dot{I}_X = -\dot{I}_Y$.

Compared to the energy flow $\dot{E}_Y$, the information flow $\dot{I}_Y$ is a more nebulous thermodynamic quantity. The information flow is a component of the total change in system free energy due to the dynamics of Y, which we call the *transduced free energy*. At steady state, Y increases the free energy $F = E - k_\mathrm{B} T S$ of the system at a rate

$$\dot{F}_Y \equiv \dot{E}_Y - k_\mathrm{B} T_Y \dot{S}_Y \tag{2.50a}$$

$$= \dot{E}_Y + k_\mathrm{B} T_Y \dot{I}_Y . \tag{2.50b}$$

$\dot{S}_Y$ is the rate at which Y increases the system entropy. At steady state the marginal entropy $S[Y] = S[X, Y] + I[X, Y] - S[X]$ is constant, so that the entropic part of the transduced free energy (quantifying changes in joint entropy) is given by the information flow [1].

To gain a better intuition, consider the following alternative definition for the information flow:

$$\dot{I}_Y \equiv \lim_{\tau \to 0} \frac{I\left[X(t), Y(t+\tau)\right] - I\left[X(t), Y(t)\right]}{\tau} . \tag{2.51}$$

Equation (2.51) shows that the information flow is the rate at which the dynamics of the Y subsystem increase the mutual information $I[X, Y]$ between the two subsystems. This mutual information is a thermodynamic resource, which can be used by, e.g., the X subsystem: if the X dynamics decrease the mutual information ($\dot{I}_X < 0$), then the subsystem second law allows X to entirely convert heat from its environment ($\dot{Q}_X > 0$) into output work ($-\dot{W}_X > 0$). If only X is observed, it appears to convert heat directly into work, seemingly a violation of the second law. Only when the information flow is taken into account, does the validity of the second law become apparent. Thus a useful way of thinking about the information flow $\dot{I}_Y$ is as the capacity of subsystem X to convert heat into work.

2.4.4 Constraints on Bipartite Engine Setups

These thermodynamic laws constrain the possible flows within bipartite engines. Figure 2.1 illustrates several example operational modes that generate output work $-\dot{W}_X$. One such example is the *conventional engine*, where free energy from input work $\dot{W}_Y$ is transduced solely in the form of energy flow between the Y and X subsystems. In this case, one subsystem takes in input work, the other produces output work, and both subsystems dissipate heat to their environments. Alternatively, it is possible to construct a pure *information engine*, where free energy is instead exchanged solely via information flow: one subsystem (here Y) takes input work, creates mutual information between Y and X, and dissipates heat $-\dot{Q}_Y$ to the environment; the other subsystem then uses the mutual information to extract heat from the environment, which it then turns into output work $-\dot{W}_X$.

Conventional and information engines represent extremes, each utilizing only one type of free-energy transduction. More generally, work converters feature both internal energy and information flows (Fig. 2.1c); we call these *hybrid* engines. Finally, bipartite machines with access to different sources of fluctuations can operate as *heat engines* (Fig. 2.1d), producing output power by leveraging a temperature difference. As discussed in Sect. 2.4.3, the directionality of internal information and energy flows is constrained [13].

The conventional and information engines defined here may at first glance bear some superficial similarities to the *power-stroke* and *Brownian-ratchet* archetypes for molecular motors [14, 15]; however, while the conventional/information engine

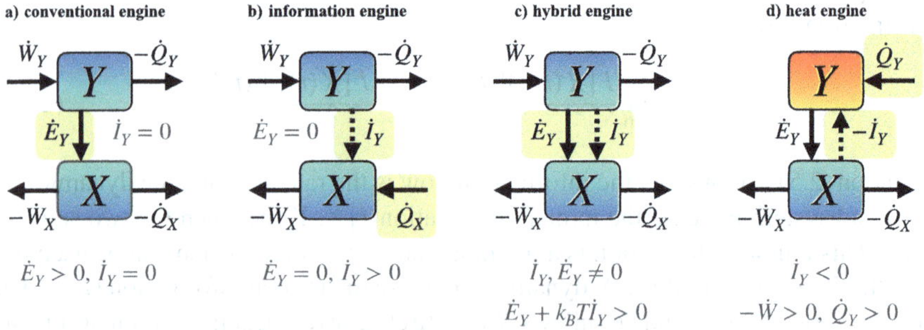

Fig. 2.1 Different operational modes of bipartite thermodynamic engines, with their defining constraints below. **a** Conventional engine. **b** Information engine. **c** Hybrid engine. **d** Heat engine ($T_Y > T_X$). Arrows show direction of energy and information flows, all symbols are positive quantities for the operational modes depicted

distinction describes how free energy is transduced within a multicomponent molecular machine, one version of the power-stroke/Brownian-ratchet distinction quantifies how free-energy differences and driving forces are split between forward and reverse rates of specific state transitions [14]. To my knowledge, there is no necessary relationship between conventional engines and such power-stroke motors, and while pure information engines and Brownian ratchets coincide under certain conditions [16], there is no known general relationship.

2.4.5 *Efficiency Metrics for Bipartite Machines*

Work Transducers

Many molecular machines transduce free energy from one form to another, taking in input work $\dot{W}_Y$ via one subsystem (Y, without loss of generality), and outputting work $-\dot{W}_X$ via the other subsystem. Examples include the conventional, information, and hybrid engines in Fig. 2.1. For work transducers, thermodynamic efficiency can be defined in a straightforward manner as the ratio of output to input work,

$$\eta_{\mathrm{T}} \equiv \frac{-\dot{W}_X}{\dot{W}_Y} \leq 1. \tag{2.52}$$

The bipartite structure allows us to resolve more fine-grained details of free-energy transduction *within* the system. Combining the first (Eq. (2.44a)) and second (Eq. (2.49)) laws of bipartite stochastic thermodynamics at steady state yields two nested inequalities relating the free-energy inputs and outputs of each subsystem [17, 18]:

$$\dot{W}_Y \geq \dot{E}_Y + k_{\mathrm{B}}T\dot{I}_Y \geq -\dot{W}_X. \tag{2.53}$$

Here the sum of the energy and information flows, the internally transduced free energy $\dot{E}_Y + k_{\mathrm{B}}T\dot{I}_Y$, acts as a bottleneck between input and output work. This motivates the introduction of subsystem efficiencies [13, 17], which quantify the efficiency of free-energy transduction for each subsystem:

$$\eta_Y \equiv \frac{\dot{E}_Y + k_{\mathrm{B}}T\dot{I}_Y}{\dot{W}_Y} \leq 1, \tag{2.54a}$$

$$\eta_X \equiv \frac{-\dot{W}_X}{\dot{E}_Y + k_{\mathrm{B}}T\dot{I}_Y} \leq 1. \tag{2.54b}$$

Their product is the thermodynamic efficiency of the whole system, $\eta_{\mathrm{T}} = \eta_Y\eta_X$. Other subsystem efficiency definitions have also been proposed to quantify the efficiency of converting information into heat energy [11].

Transport Motors

Many molecular machines, rather than transducing energy from one form to another, consume free energy to transport cargo from one place to another. Examples include transport motors such as kinesin pulling a diffusive cargo, and the flagellar motor pushing a cell. For these kinds of systems, we can think of the motor and cargo as the Y and X subsystems, respectively. In this case, the output of the motor is motion of the cargo at some average velocity $\langle v_X \rangle$ against the force of viscous friction, for Stokes flow equal to the friction coefficient ζ_X times the velocity. The average work rate against this viscous friction force is thus $\zeta_X \langle v_X \rangle^2$, suggesting the definition of the *Stokes efficiency* [19]:

$$\eta_S \equiv \frac{\zeta_X \langle v_X \rangle^2}{\dot{W}_Y} \leq 1. \tag{2.55}$$

For transport systems, the motor itself still has a well-defined subsystem efficiency (Eq. (2.54a)), quantifying how much of the input work is made available to the cargo as transduced free energy [20].

2.5 Many Degrees of Freedom: Multipartite Systems

While Sect. 2.4 focuses on bipartite systems, many of the main results generalize to multipartite systems with $N > 2$ degrees of freedom. As with the bipartite assumption, a system is multipartite when all subsystems are influenced by independent fluctuations, with the same mathematical consequences as in the bipartite case. Many molecular machines of interest are well-described as multipartite systems, for example collective transport systems where anywhere from one to over a hundred transport motors jointly pull a single cargo [21, 22], or the bacterial flagellar motor where ≈ 10 stators collectively apply torque to turn a single rotor.

Consider an overdamped stochastic system with N degrees of freedom, denoted $\boldsymbol{x} = \{x_i\}_{i=1}^N$, each in contact with a heat bath at temperature T_i. The system is characterized by a mobility tensor $\boldsymbol{\mu}$, which is related to the hydrodynamic friction tensor $\boldsymbol{\zeta}$ by $\boldsymbol{\mu} = \boldsymbol{\zeta}^{-1}$ [23, 24], as well as a diffusion tensor $\boldsymbol{D}$. For multipartite systems the mobility, friction, and diffusion tensors are all diagonal, so that each coordinate is subject to independent thermal noise [25]. I assume their elements to be independent of time and position. D_i, ζ_i, and μ_i denote the coefficients of the ith subsystem.

The system dynamics depend on the forces that influence it; these forces may either result from a conservative potential $V(\boldsymbol{x}, t)$, or be nonconservative forces $\boldsymbol{f}_{\text{nc}}(\boldsymbol{x}, t)$ which do not arise from a potential. Here I also allow both conservative and nonconservative forces to in general be time-dependent due to external control. The total force vector acting on the system is $\boldsymbol{f}(\boldsymbol{x}, t) = \boldsymbol{f}_{\text{nc}}(\boldsymbol{x}, t) - \nabla V(\boldsymbol{x}, t)$.

The system evolves dynamically according to a multidimensional overdamped Langevin equation,

$$\dot{\boldsymbol{x}} = \boldsymbol{\mu} \boldsymbol{f}(\boldsymbol{x}, t) + \boldsymbol{\xi}(t), \tag{2.56}$$

where the random fluctuations $\boldsymbol{\xi}(t)$ have zero mean and correlations satisfying

$$\langle \boldsymbol{\xi}(t) \boldsymbol{\xi}^{\top}(t') \rangle = 2\boldsymbol{D}\, \delta(t - t'). \tag{2.57}$$

I assume the fluctuation-dissipation theorem, so that elements of the diffusion and mobility tensors are related by [26, 27]

$$D_{ii} = \mu_{ii}\, k_{\mathrm{B}} T_i, \tag{2.58}$$

for Boltzmann's constant k_{B}.

These dynamics can equivalently be described at the level of the probability distribution $p(\boldsymbol{x}, t)$ by a Fokker-Planck equation (FPE):

$$\frac{\partial}{\partial t} p(\boldsymbol{x}, t) = -\nabla \cdot \boldsymbol{J}(\boldsymbol{x}, t), \tag{2.59}$$

with probability flux vector

$$\boldsymbol{J}(\boldsymbol{x}, t) \equiv \boldsymbol{\mu}\, \boldsymbol{f}(\boldsymbol{x}, t)\, p(\boldsymbol{x}, t) - \boldsymbol{D}\, \nabla p(\boldsymbol{x}, t). \tag{2.60}$$

Following the same steps as for bipartite systems, the ensemble-averaged energy flow rates are

$$\dot{E}_i = \left\langle \frac{\partial V}{\partial x_i} \frac{J_i(\boldsymbol{x}, t)}{p(\boldsymbol{x}, t)} \right\rangle, \tag{2.61a}$$

$$\dot{W}_i = \left\langle f_{\mathrm{nc}}(\boldsymbol{x}, t) \frac{J_i(\boldsymbol{x}, t)}{p(\boldsymbol{x}, t)} \right\rangle, \tag{2.61b}$$

$$\dot{Q}_i = \left\langle -f(\boldsymbol{x}, t) \frac{J_i(\boldsymbol{x}, t)}{p(\boldsymbol{x}, t)} \right\rangle. \tag{2.61c}$$

For each subsystem, these flows satisfy a local first law:

$$\dot{E}_i = \dot{W}_i + \dot{Q}_i. \tag{2.62}$$

Each subsystem likewise satisfies a second law [25],

$$\dot{\Sigma}_i = \dot{S}_i[X] - \dot{Q}_i / (k_{\mathrm{B}} T_i) \geq 0, \tag{2.63}$$

for entropy flow

$$\dot{S}_i[X] = \left\langle -\frac{\partial}{\partial x_i} \ln p(\boldsymbol{x}, t) \frac{J_i(\boldsymbol{x}, t)}{p(\boldsymbol{x}, t)} \right\rangle, \tag{2.64}$$

and entropy production rate

$$\dot{\Sigma}_i = D_i^{-1} \left\langle \left[\frac{J_i(\boldsymbol{x}, t)}{p(\boldsymbol{x}, t)} \right]^2 \right\rangle. \tag{2.65}$$

Due to the multipartite assumption, the total entropy production rate is simply the sum of all subsystem entropy production rates [25],

$$\dot{\Sigma} = \sum_{i=1}^{N} \dot{\Sigma}_i. \tag{2.66}$$

As for bipartite systems, the system entropy rate $\dot{S}_X[\boldsymbol{X}]$ can be decomposed into a marginal entropy term and an information-theoretic term,

$$\dot{S}_X[\boldsymbol{X}] = \left\langle -\frac{\partial}{\partial x_i} \ln p(\boldsymbol{x}, t) \frac{J_i(\boldsymbol{x}, t)}{p(\boldsymbol{x}, t)} \right\rangle \tag{2.67a}$$

$$= \left\langle -\frac{\mathrm{d}}{\mathrm{d}x_i} \ln p_{X_i}(x_i, t) \frac{J_i(\boldsymbol{x}, t)}{p(\boldsymbol{x}, t)} \right\rangle - \left\langle \frac{\partial}{\partial x_i} \ln \frac{p(\boldsymbol{x}, t)}{\Pi_{j=1}^{N} p_{X_j}(x_j, t)} \frac{J_i(\boldsymbol{x}, t)}{p(\boldsymbol{x}, t)} \right\rangle \tag{2.67b}$$

$$= \mathrm{d}_t S[X_i] - \dot{I}_i. \tag{2.67c}$$

Here the information flow $\dot{I}_i$ is the rate at which the dynamics of the i'th subsystem change the total correlation (Eq. (2.9)), which is one possible multivariate generalization of the mutual information. The subsystem second laws then become

$$\dot{\Sigma}_i = \mathrm{d}_t S_{X_i}[X_i] - \dot{Q}_i / (k_{\mathrm{B}} T_i) - \dot{I}_i \geq 0 \tag{2.68}$$

The appearance in the second law of the total correlation shows that it is the correct multivariate generalization of the mutual information as a thermodynamic resource.

While for bipartite systems the energy and information flows can be described as flows from one system to another, this description is much less straightforward for multipartite systems. In general such a description can only be assigned when the graph describing interactions between subsystems is a tree [28]. This is the case for the collective transport system considered in Chap. 4, where each of N motors can only interact with the sole cargo, and thus the energy flows for each motor can be considered flows from motor to cargo. Otherwise, in this thesis I will avoid delving into these fraught accounting considerations, and leave it for future work.

2.6 Other Generalizations

While my focus in this thesis is on analysis of single molecular machines, or single collections of linked machines, when large collections are considered together they can be modeled using deterministic chemical reaction networks, with state concentrations replacing state probabilities as the relevant variables. Nonequilibrium steady-state thermodynamics can still be formulated for chemical reaction networks [29], and when these networks are bipartite the energy and information flows can be defined as for single molecular machines [16]. This formalism has frequently been used to study the thermodynamics of synthetic molecular machines [30, 31].

Efforts have also been made to extend certain results to systems lacking the bipartite structure [13, 32, 33]. I will derive certain results for non-bipartite/multipartite systems in Chaps. 3 and 8.

2.7 Bipartite Models for Molecular Machines

To give some context for how others have used stochastic thermodynamics to study multicomponent molecular machines, here I highlight several examples of bipartite models for molecular machines including ATP synthase, kinesin, and a synthetic rotary motor (see Fig. 2.2). The lens of bipartite stochastic thermodynamics leads to new kinds of analysis in these systems.

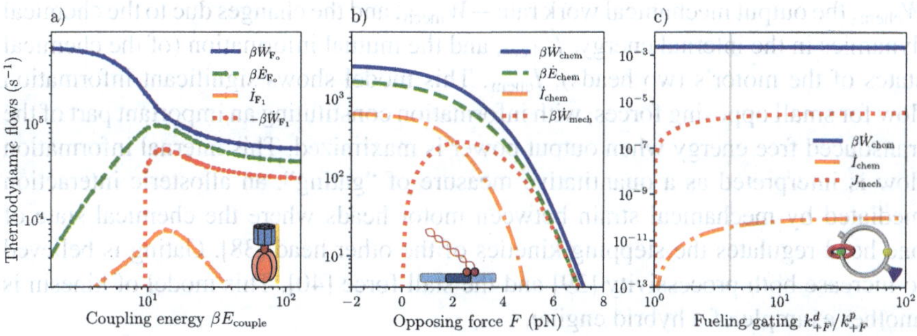

Fig. 2.2 Insights from analysis of the energy and information flows in simple models of molecular machines. **a** Energy and information flows for F_O and F_1 in ATP synthase, as functions of the coupling energy βE_{couple}, from the model considered in Ref. [18]. **b** Energy and information flows from the chemical and mechanical components of the kinesin motor model studied in Ref. [36], as functions of the opposing force F. **c** Chemical-work consumption, information flow, and mechanical flux in a model, as functions of the fuelling gating, for the synthetic rotary motor studied in Ref. [37]. Gray vertical line indicates experimental conditions. Panel (**a**) adapted with permission from Ref. [18]; copyrighted by the American Physical Society. Inset schematic and code for (**c**) adapted from Ref. [37] (CC BY 4.0)

The coupled rotary motors of ATP synthase constitute a textbook example of a bipartite molecular machine. The rotational degrees of freedom in ATP synthase, the c-ring in F_o and the γ-shaft in F_1, can be modelled using overdamped Langevin equations coupled by a rotationally symmetric joint potential energy [18, 34, 35]. One of the most notable findings [35] is that output power depends non-monotonically on the coupling strength between F_o and F_1. While efficiency is highest for tight coupling, the output power is maximized at intermediate coupling strength (Fig. 2.2a). This somewhat counterintuitive result can be understood by considering the dominant transition paths (inchworming and slippage) during coupled rotation of the two motors [35]: at low coupling strength the motors are decoupled, while at high coupling strength they can only rotate in lockstep; intermediate coupling strength allows enough flexibility for F_o and F_1 to separately cross their respective energy barriers (inchworming) while keeping slippage minimal.

Exploring a more detailed picture of energy and information flows sheds further light on the inner workings of ATP synthase. Figure 2.2a shows the input and output work rates $\dot{W}_{F_o}$ and $-\dot{W}_{F_1}$, along with the energy and information flows $\dot{E}_{F_o}$ and $\dot{I}_{F_1}$. As with the output power, the transduced free energy and energy flow from F_o to F_1 are both maximized at intermediate coupling, along with the information flow. This information flow from F_1 to F_o allows F_o to decrease the heat it dissipates to the environment, thereby increasing the energy transduced to F_1 [18]. This model of ATP synthase is an example of a hybrid engine.

Reference [36] considered flows of energy and information between the heads of individual kinesin and myosin motors, as well as changes in energy and information due to contributions from chemical and mechanical dynamics. Figure 2.2b illustrates the thermodynamics of kinesin stepping. Shown are the input chemical work rate $\dot{W}_{\text{chem}}$, the output mechanical work rate $-\dot{W}_{\text{mech}}$, and the changes due to the chemical dynamics in the internal energy, $\dot{E}_{\text{chem}}$, and the mutual information (of the chemical states of the motor's two heads), $\dot{I}_{\text{chem}}$. This model shows significant information flow for small opposing forces, with information constituting an important part of the transduced free energy when output power is maximized. This internal information flow is interpreted as a quantitative measure of "gating", an allosteric interaction mediated by mechanical strain between motor heads where the chemical state of one head regulates the stepping kinetics of the other head [38]. Gating is believed to increase both processivity [39] and the stall force [40]. This model of kinesin is another example of a hybrid engine.

The first nanoscale autonomous chemically fueled molecular motor [41] was analyzed from a thermodynamic standpoint in Ref. [30]. This rotary motor (Fig. 1.1a fourth panel) consists of two interlocking molecular rings with two "docking stations" where motion of the smaller ring is alternately blocked or permitted by binding/unbinding of chemical fuel. The motor was modelled as bipartite, with coupled chemical and mechanical degrees of freedom. In this case, the experimentally synthesized molecular motor was shown to operate as an information ratchet, with free energy transduced entirely via information flow $\dot{I}_{\text{chem}}$ (Fig. 2.2c). While this motor is strikingly less efficient than biological molecular machines, with the chemical subsystem's efficiency on the order of 10^{-8}, an exploration in parameter space around the physically realized model allowed the authors to determine design principles for future improvements. For example, by adjusting the "fuelling gating" (the ratio of two rate constants, quantifying attachment bias of the chemical fuel), the mechanical flux J_{mech} can be increased at constant chemical input work $\dot{W}_{\text{chem}}$ (Fig. 2.2c). This increase in performance occurs via a corresponding increase in the information flow, and thus the subsystem efficiency (Eq. (2.54a)) of free-energy transduction from the chemical to mechanical subsystems. This synthetic molecular machine is an example of an information engine.

2.8 Thermodynamic Inference

Many thermodynamic quantities of interest, for example efficiencies and internal energy and information flows, are difficult to quantify experimentally. Calculating these quantities generally requires full knowledge of the joint nonequilibrium probability distribution governing the relevant degrees of freedom, along with knowledge of all conservative and nonconservative forces acting on the system. Such information is typically beyond our current ability to measure for systems as complex as biological molecular machines, and can be difficult to compute even for numerical simulations of model systems. As such, we turn to thermodynamic inference: the study, development, and application of mathematical tools to infer hidden thermodynamic properties from quantities that can be observed experimentally [42].

Thermodynamic inference is often accomplished via the use of bounds on entropy production. One generic example is bounding the steady-state thermodynamic efficiency of an isothermal molecular machine converting between different forms of work. Typical single-molecule experiments probe motor proteins working against applied load forces, and measure the output work rate $\dot{W}_{\text{out}}$ but not the input work rate $\dot{W}_{\text{in}}$. In this case, a lower bound $\dot{\Sigma}_{\text{LB}}$ on the entropy production rate leads to an upper bound on the thermodynamic efficiency, $\eta_{\text{T}} \leq 1/\left(1 + \dot{\Sigma}_{\text{LB}}/\dot{W}_{\text{out}}\right)$. An important goal in stochastic thermodynamics is then to derive entropy production bounds, calculable from available data, which can be inserted into relationships such as these.

2.8.1 Thermodynamic Uncertainty Relations

A widely used entropy production bound emerging from stochastic thermodynamics is the family of thermodynamic uncertainty relations (TURs) [43]. These inequalities relate entropy production to the means and variances of time-integrated dynamic or thermodynamic currents. TURs typically take the form

$$\Sigma_\tau \geq \frac{2\langle J_\tau\rangle}{\text{Var}\,(J_\tau)}, \tag{2.69}$$

where Σ_τ is the entropy production over a time interval τ, and J_τ is any suitably defined current. In the short-time limit, bounds on the entropy production rate are obtained.

An immediate application of the TUR, first demonstrated in Ref. [44], is inferring the efficiencies of molecular transport motors from measurements of their average velocity $\langle v \rangle$ and effective diffusivity

$$D_{\mathrm{eff}} \equiv \lim_{t\to\infty} \frac{\langle \Delta x^2 \rangle - \langle \Delta x \rangle^2}{2t}. \tag{2.70}$$

In particular, the thermodynamic efficiency (Eq. (2.52)) of a motor pulling against a constant load force f, and the Stokes efficiency (Eq. (2.55)) of a motor pulling a diffusive cargo (with diffusion coefficient D_{c}) against viscous drag, are upper bounded by

$$\eta_{\mathrm{T}} \le \left(1 + \frac{\langle v \rangle}{\beta D_{\mathrm{eff}} f_{\mathrm{load}}}\right)^{-1}, \tag{2.71a}$$

$$\eta_{\mathrm{S}} \le \frac{D_{\mathrm{eff}}}{D_{\mathrm{c}}}. \tag{2.71b}$$

The TUR can also be used to quantitatively assess the performance of molecular machines, by measuring how close their true entropy production rate comes to saturating the bound. This has been done, for example, for a simulated artificial molecular motor, revealing significant deviation from the bound [45]. Investigating the source of these inefficiencies can uncover design principles for improving performance.

Beyond the canonical form of the TUR (Eq. (2.69)), other related bounds include TUR-type bounds on subsystem entropy production rates valid at short times [46], and a pair of bipartite TUR bounds for subsystem entropy production rates that hold more generally [47].

2.8.2 Other Bounds on Entropy Production

While the TUR has seen applications towards inference for molecular machines, many other bounds on entropy production have been derived through the formalism of stochastic thermodynamics. Most prominently, these include a collection of *thermodynamic speed limits* [48–51] that lower bound the entropy production in terms of the time taken to move a system from one probability distribution to another.

While constraints on entropy production typically provide lower bounds, Dechant [52] recently derived an upper bound on the entropy production rate of a stochastic system, $\dot{\Sigma} \le \beta \langle \boldsymbol{f}_{\mathrm{nc}}^{\mathsf{T}} \boldsymbol{\mu} \boldsymbol{f}_{\mathrm{nc}} \rangle$, in terms of an ensemble-averaged norm of the non-conservative forces $\boldsymbol{f}_{\mathrm{nc}}$. Intuitively, the magnitude of nonconservative forces limits how far from equilibrium the system can be driven, thus limiting the entropy production rate. Combined with previously derived lower bounds on entropy production, upper bounds hold promise for more precise inference.

Entropy production can also be inferred directly, for example using the *Variance Sum Rule* (VSR) [53, 54], which requires measurements of the second time derivative of the variance of position displacement along with the variance of the forces acting on the system. While the VSR has yet to be applied to the study of molecular machines, I believe it has significant potential as an inference tool.

In the next chapter, I will derive a new entropy production bound that will prove invaluable for studying multicomponent molecular machines.

References

1. Ehrich J, Sivak D (2023) Energy and information flows in autonomous systems. Front Phys 11:1108357
2. Cover TM, Thomas JA (2006) Elements of information theory, 2nd edn. Wiley-Interscience, Hoboken, NJ
3. Crooks GE (2017) On measures of entropy and information. Tech Note 9(4)
4. Seifert U (2012) Stochastic thermodynamics, fluctuation theorems and molecular machines. Rep Prog Phys 75(12):126001
5. Einstein A et al (1905) On the motion of small particles suspended in liquids at rest required by the molecular-kinetic theory of heat. Annalen der physik 17(549–560):208
6. Van Kampen NG (1981) Itô versus Stratonovich. J Stat Phys 24:175–187
7. Ehrich J (2020) Coupled and hidden degrees of freedom in stochastic thermodynamics. arXiv:2007.15223
8. Seifert U (2005) Entropy production along a stochastic trajectory and an integral fluctuation theorem. Phys Rev Lett 95(4):040602
9. Hartich D, Barato AC, Seifert U (2014) Stochastic thermodynamics of bipartite systems: transfer entropy inequalities and a Maxwell's demon interpretation. J Stat Mech Theory Exp 2:P02016
10. Diana G, Esposito M (2014) Mutual entropy production in bipartite systems. J Stat Mech Theory Exp 2014:P04010
11. Horowitz JM, Esposito M (2014) Thermodynamics with continuous information flow. Phys Rev X 4(3):031015
12. Brittain RA, Jones NS, Ouldridge TE (2017) What we learn from the learning rate. J Stat Mech Theory Exp 2017(6):063502
13. Leighton MP, Ehrich J, Sivak DA (2024) Information arbitrage in bipartite heat engines. Phys Rev X 14:041038
14. Wagoner JA, Dill KA (2016) Molecular motors: power strokes outperform Brownian ratchets. J Phys Chem B 120(26):6327–6336
15. Hwang W, Karplus M (2019) Structural basis for power stroke vs. Brownian ratchet mechanisms of motor proteins. Proc Natl Acad Sci 116(40):19777–19785
16. Penocchio E, Avanzini F, Esposito M (2022) Information thermodynamics for deterministic chemical reaction networks. J Chem Phys 157(3)
17. Barato AC, Seifert U (2017) Thermodynamic cost of external control. New J Phys 19(7):073021
18. Lathouwers E, Sivak DA (2022) Internal energy and information flows mediate input and output power in bipartite molecular machines. Phys Rev E 105(2):024136
19. Wang H, Oster G (2002) The Stokes efficiency for molecular motors and its applications. EPL (Eur Lett) 57(1):134
20. Leighton MP, Sivak DA (2023) Inferring subsystem efficiencies in bipartite molecular machines. Phys Rev Lett 130(17):178401

21. Leighton MP, Sivak DA (2022) Performance scaling and trade-offs for collective motor-driven transport. New J Phys 24(1):013009
22. Leighton MP, Sivak DA (2022) Dynamic and thermodynamic bounds for collective motor-driven transport. Phys Rev Lett 129(11):118102
23. Bogdan Cichocki B, Ubbo Felderhof K, Hinsen EW, Blawzdziewicz J (1994) Friction and mobility of many spheres in stokes flow. J Chem Phys 100(5):3780–3790
24. Cichocki B, Jones RB, Kutteh R, Wajnryb E (2000) Friction and mobility for colloidal spheres in stokes flow near a boundary: the multipole method and applications. J Chem Phys 112(5):2548–2561
25. Horowitz JM (2015) Multipartite information flow for multiple Maxwell demons. J Stat Mech Theory Exp 2015(3):P03006
26. Maes C, Netočný K, Wynants B (2008) Steady state statistics of driven diffusions. Phys A Stat Mech Appl 387(12):2675–2689
27. Ito S, Dechant A (2020) Stochastic time evolution, information geometry, and the cramér-rao bound. Phys Rev X 10(2):021056
28. Leighton MP, Sivak DA Accounting for energy and information flows in multicomponent stochastic systems, in prep
29. Rao R, Esposito M (2016) Nonequilibrium thermodynamics of chemical reaction networks: wisdom from stochastic thermodynamics. Phys Rev X 6(4):041064
30. Amano S, Esposito M, Kreidt E, Leigh DA, Penocchio E, Roberts BMW (2022) Insights from an information thermodynamics analysis of a synthetic molecular motor. Nat Chem 14(5):530–537
31. Corra S, Bakić MT, Groppi J, Baroncini M, Silvi S, Penocchio E, Esposito M, Credi A (2022) Kinetic and energetic insights into the dissipative non-equilibrium operation of an autonomous light-powered supramolecular pump. Nat Nanotechnol 17(7):746–751
32. Chetrite R, Rosinberg ML, Sagawa T, Tarjus G (2019) Information thermodynamics for interacting stochastic systems without bipartite structure. J Stat Mech Theory Exp 11:114002
33. Leighton MP, Sivak DA (2024) Jensen bound for the entropy production rate in stochastic thermodynamics. Phys Rev E 109(1):L012101
34. Kei-ichi O, Gerhard H (2015) Elasticity, friction, and pathway of γ-subunit rotation in F_oF_1-ATP synthase. Proc Natl Acad Sci 112(34):10720–10725
35. Lathouwers E, Lucero JNE, Sivak DA (2020) Nonequilibrium energy transduction in stochastic strongly coupled rotary motors. J Phys Chem Lett 11(13):5273–5278
36. Takaki R, Mugnai ML, Thirumalai D (2022) Information flow, gating, and energetics in dimeric molecular motors. Proc Natl Acad Sci 119:e2208083119
37. Amano S, Esposito M, Kreidt E, Leigh DA, Penocchio E, Roberts BMW (2021) Insights from an information thermodynamics analysis of a synthetic molecular motor. ChemRxiv
38. Yildiz A, Tomishige M, Gennerich A, Vale RD (2008) Intramolecular strain coordinates kinesin stepping behavior along microtubules. Cell 134(6):1030–1041
39. Andreasson JOL, Milic B, Chen G-Y, Guydosh NR, Hancock WO, Block SM (2015) Examining kinesin processivity within a general gating framework. Elife 4:e07403
40. Hinczewski M, Tehver R, Thirumalai D (2013) Design principles governing the motility of myosin V. Proc Natl Acad Sci 110(43):E4059–E4068
41. Wilson MR, Solà J, Carlone A, Goldup SM, Lebrasseur N, Leigh DA (2016) An autonomous chemically fuelled small-molecule motor. Nature 534(7606):235–240
42. Seifert U (2019) From stochastic thermodynamics to thermodynamic inference. Annu Rev Condens Matter Phys 10:171–192
43. Horowitz JM, Gingrich TR (2020) Thermodynamic uncertainty relations constrain non-equilibrium fluctuations. Nat Phys 16(1):15–20
44. Pietzonka P, Barato AC, Seifert U (2016) Universal bound on the efficiency of molecular motors. J Stat Mech Theory Exp 12:124004
45. Albaugh A, Rueih-Sheng F, Geyao G, Gingrich TR (2023) Limits on the precision of catenane molecular motors: Insights from thermodynamics and molecular dynamics simulations. J Chem Theory Comput 20(1):1–6

46. Otsubo S, Ito S, Dechant A, Sagawa T (2020) Estimating entropy production by machine learning of short-time fluctuating currents. Phys Rev E 101(6):062106
47. Tanogami T, Van Tan V, Saito K (2023) Universal bounds on the performance of information-thermodynamic engine. Phys Rev Res 5(4):043280
48. Shiraishi N, Funo K, Saito K (2018) Speed limit for classical stochastic processes. Phys Rev Lett 121(7):070601
49. Falasco G, Esposito M (2020) Dissipation-time uncertainty relation. Phys Rev Lett 125(12):120604
50. Van Tan V, Hasegawa Y et al (2020) Unified approach to classical speed limit and thermodynamic uncertainty relation. Phys Rev E 102(6):062132
51. Nakazato M, Ito S (2021) Geometrical aspects of entropy production in stochastic thermodynamics based on Wasserstein distance. Phys Rev Res 3(4):043093
52. Dechant A (2023) Upper bounds on entropy production in diffusive dynamics. arXiv:2310.17929
53. Di Terlizzi I, Gironella M, Herráez-Aguilar D, Timo Betz F, Monroy MB, Ritort F (2024) Variance sum rule for entropy production. Science 383(6686):971–976
54. Di Terlizzi I, Baiesi M, Ritort F (2024b) Variance sum rule: proofs and solvable models. New J Phys

Chapter 3
Jensen Bound on the Entropy Production Rate for Multicomponent Stochastic Systems

Bounding and estimating entropy production has long been an important goal of nonequilibrium thermodynamics. I derive a lower bound on the total and subsystem entropy production rates of continuous stochastic systems. I first derive this 'Jensen bound' for multipartite overdamped Langevin dynamics, then consider several extensions, allowing for position-dependent diffusion coefficients, underdamped dynamics, and non-multipartite overdamped dynamics.

3.1 Introduction

Bounds on entropy production are useful tools in thermodynamics, placing constraints on the space of physically realizable systems and processes. For an example, we need look no further than the second law itself. The second law was famously used by Carnot to bound the efficiency of heat engines [1], leading to improved steam-engine designs with greater efficiencies. Entropy-production bounds likewise serve as tools for thermodynamic inference [2], where observations of a system's dynamical behaviour are used to constrain hidden thermodynamic details.

The last decade has seen a flurry of entropy-production bounds derived through the framework of stochastic thermodynamics [3], the most well-known being the thermodynamic uncertainty relation (TUR). First conjectured in Ref. [4] and later proven more generally in Ref. [5], the TUR encompasses a family of relations bounding entropy production using fluctuations of observable currents (Ref. [6] reviews recent developments). Important applications include bounding and inferring the efficiency of motor proteins from experimental data [7]. Other stochastic-thermodynamic entropy-production bounds include geometrical bounds based on the Wasserstein distance [8] and various thermodynamic speed limits [9–11].

In this chapter, I derive a new lower bound on entropy production rates for multicomponent stochastic systems which I call the Jensen bound, providing a general derivation for continuous stochastic systems with multiple degrees of freedom. In

M. Leighton, *Stochastic Thermodynamics of Multicomponent Molecular Machines*, Springer Theses, https://doi.org/10.1007/978-3-032-02204-2_3

particular, I focus on four main objectives. I first derive the Jensen bound in full generality for multipartite overdamped Langevin dynamics, allowing for external control, different temperatures, and different mean coordinate rates of change. I then establish three additional extensions to the Jensen bound: inhomogeneous diffusion coefficients, underdamped dynamics, and non-multipartite overdamped dynamics. I also explore the relationship between the Jensen bound and the TUR, showing explicitly how for some classes of systems the Jensen bound can be derived from short-time limits of different TUR formulations.

3.2 Multipartite Overdamped Langevin Dynamics

Consider a multipartite stochastic system with N degrees of freedom evolving according to overdamped Langevin dynamics, as characterized in Sect. 2.5. As shown in Sect. 2.5, the entropy production rate of the ith subsystem is

$$\dot{\Sigma}_i = D_i^{-1}\left\langle\left[\frac{J_i(\boldsymbol{x},t)}{p(\boldsymbol{x},t)}\right]^2\right\rangle. \tag{3.1}$$

The total entropy production rate is the sum of all subsystem entropy production rates,

$$\dot{\Sigma} = \sum_{i=1}^{N}\dot{\Sigma}_i. \tag{3.2}$$

Given the functional form of Eq. (3.1), Jensen's inequality [12] dictates that

$$\left\langle\left[\frac{J_i(\boldsymbol{x},t)}{p(\boldsymbol{x},t)}\right]^2\right\rangle \geq \left\langle\frac{J_i(\boldsymbol{x},t)}{p(\boldsymbol{x},t)}\right\rangle^2. \tag{3.3}$$

On the righthand side I identify the mean rate of change $\langle\dot{x}_i\rangle$ of the ith coordinate, which is equal to the ensemble-average of the probability current divided by the probability distribution [3]:

$$\langle\dot{x}_i\rangle = \left\langle\frac{J_i(\boldsymbol{x},t)}{p(\boldsymbol{x},t)}\right\rangle. \tag{3.4}$$

Two inequalities follow from this, respectively constraining the partial and total entropy production rates:

$$\dot{\Sigma}_i \geq D_i^{-1}\langle\dot{x}_i\rangle^2, \tag{3.5a}$$

$$\dot{\Sigma} \geq \sum_{i=1}^{N} D_i^{-1}\langle\dot{x}_i\rangle^2. \tag{3.5b}$$

I call Eq. (3.5a) the *subsystem Jensen bound* and Eq. (3.5b) the *total Jensen bound.* These can equivalently be written in terms of friction coefficients rather than diffusion coefficients.

Note that this derivation allows for time-dependent forces, different subsystem temperatures, and different mean coordinate rates of change. In the following sections, I will generalize the Jensen bound in three different directions to allow for position-dependent diffusion coefficients, underdamped dynamics, and non-multipartite dynamics.

3.3 Position-Dependent Diffusion Coefficients

Suppose now that the system has position-dependent diffusion coefficients, such that $D_i = D_i(\boldsymbol{x})$. By the fluctuation-dissipation theorem (2.58), the mobility and friction coefficients thus also depend on position. The entropy production rate for the ith subsystem is then [13]

$$\dot{\Sigma}_i = \left\langle D_i(\boldsymbol{x})^{-1} \left[\frac{J_i(\boldsymbol{x},t)}{p(\boldsymbol{x},t)}\right]^2\right\rangle . \tag{3.6}$$

To bound this quantity, I use the Cauchy-Schwarz inequality, which applied to the covariance of two random variables X and Y gives [14]

$$\langle XY\rangle^2 \leq \langle X^2\rangle\langle Y^2\rangle . \tag{3.7}$$

Specializing to $X = J_i(\boldsymbol{x},t)/[\sqrt{D_i(\boldsymbol{x})}p(\boldsymbol{x},t)]$ and $Y = \sqrt{D_i(\boldsymbol{x})}$ then gives

$$\left\langle\frac{J_i(\boldsymbol{x},t)}{p(\boldsymbol{x},t)}\right\rangle^2 \leq \left\langle D_i(\boldsymbol{x})^{-1} \left[\frac{J_i(\boldsymbol{x},t)}{p(\boldsymbol{x},t)}\right]^2\right\rangle \langle D_i(\boldsymbol{x})\rangle . \tag{3.8}$$

Identifying the entropy production rate $\dot{\Sigma}_i$ (3.6) on the righthand side and mean coordinate rate of change $\langle\dot{x}_i\rangle$ (3.4) on the left hand side, dividing both sides by $\langle D_i(\boldsymbol{x})\rangle$ then yields the subsystem Jensen bound

$$\dot{\Sigma}_i \geq \langle D_i(\boldsymbol{x})\rangle^{-1}\langle\dot{x}_i\rangle^2 , \tag{3.9}$$

where the constant diffusion coefficient of Eq. (3.5a) has been replaced by the steady-state average diffusion coefficient $\langle D_i(\boldsymbol{x})\rangle$. Summing the subsystem Jensen bounds for all N subsystems then yields the total Jensen bound:

$$\dot{\Sigma} \geq \sum_{i=1}^{N} \langle D_i(\boldsymbol{x}) \rangle^{-1} \langle \dot{x}_i \rangle^2. \tag{3.10}$$

While these expressions require knowledge of the nonequilibrium probability distribution to evaluate the quantities $\langle D_i(\boldsymbol{x}) \rangle$, the entropy production rates can be further lower bounded using the maximum values of the diffusion coefficients,

$$D_i^{\max} \equiv \max \{ D_i(\boldsymbol{x}) \; : \; \boldsymbol{x} \in \mathrm{dom}(\boldsymbol{x}) \} . \tag{3.11}$$

Here $\mathrm{dom}(\boldsymbol{x})$ is the domain of $\boldsymbol{x}$ over which the function $D_i(\boldsymbol{x})$ is defined. Since $D_i^{\max} \geq D_i(\boldsymbol{x})$ for all $\boldsymbol{x} \in \mathrm{dom}(\boldsymbol{x})$, we have

$$\dot{\Sigma}_i \geq \left(D_i^{\max} \right)^{-1} \langle \dot{x}_i \rangle^2, \tag{3.12a}$$

$$\dot{\Sigma} \geq \sum_{i=1}^{N} \left(D_i^{\max} \right)^{-1} \langle \dot{x}_i \rangle^2. \tag{3.12b}$$

3.4 Alternative Derivation from the Short-Time TUR

The subsystem Jensen bounds (3.5a) and (3.9) can also be derived from the short-time thermodynamic uncertainty relation. In Ref. [15], it was shown that for overdamped Langevin dynamics the entropy production rate of the ith subsystem is lower bounded by

$$\dot{\Sigma}_i \geq \frac{2 \left\langle j_{d_i} \right\rangle^2}{\mathrm{Var}\left(j_{d_i} \right) \mathrm{d}t}, \tag{3.13}$$

for any current

$$j_{d_i} \mathrm{d}t = d_i(\boldsymbol{x}, t) \circ \mathrm{d}x_i(t) \tag{3.14}$$

in the limit as $\mathrm{d}t \to 0$. Setting $d_i(\boldsymbol{x}, t) = 1$ and computing the mean and variance of the short-time current following the methods of Ref. [15] gives $\left\langle j_{d_i} \right\rangle = \langle \dot{x}_i \rangle$ and $\mathrm{Var}\left(j_{d_i}^{\tau} \right) \mathrm{d}t = 2 \langle D_i \rangle$. Inserting these identities into the short-time TUR (3.13) yields

$$\dot{\Sigma}_i \geq \langle D_i(\boldsymbol{x}) \rangle^{-1} \langle \dot{x}_i \rangle^2. \tag{3.15}$$

This is identical to the subsystem Jensen bound for multipartite overdamped Langevin dynamics with position-dependent diffusion coefficients (3.9), and simplifies to Eq. (3.5a) for constant diffusion coefficients.

3.5 Underdamped Langevin Dynamics

I now turn to underdamped Langevin dynamics, for which the equations of motion are [16]

$$\dot{\boldsymbol{x}} = \boldsymbol{v}, \tag{3.16a}$$

$$\underline{\underline{m}}\dot{\boldsymbol{v}} = -\boldsymbol{\zeta}\boldsymbol{v} + \boldsymbol{f}(\boldsymbol{x}, t) + \boldsymbol{\xi}(t). \tag{3.16b}$$

Here the random noise $\boldsymbol{\xi}(t)$ directly affects the velocity dynamics rather than position dynamics. $\boldsymbol{\xi}(t)$ is Gaussian and satisfies

$$\langle \boldsymbol{\xi}(t) \rangle = 0, \tag{3.17a}$$

$$\langle \xi_i(t)\xi_j(t') \rangle = 2k_{\mathrm{B}}T_i\, \zeta_{ij}\, \delta_{ij}\, \delta(t - t'). \tag{3.17b}$$

The entropy production rate is [16, 17]

$$\dot{\Sigma} = \sum_{i=1} \underbrace{\left\langle \frac{m_i^2}{T_i \zeta_i} \left[\frac{J_{v_i}^{\mathrm{irr}}(\boldsymbol{x}, \boldsymbol{v}, t)}{p(\boldsymbol{x}, \boldsymbol{v}, t)} \right]^2 \right\rangle}_{\dot{\Sigma}_i}, \tag{3.18}$$

for irreversible current

$$J_{v_i}^{\mathrm{irr}}(\boldsymbol{x}, \boldsymbol{v}, t) = \frac{1}{m_i} \left(-\zeta_i v_i - \frac{k_{\mathrm{B}} T_i \zeta_i}{m_i} \frac{\partial}{\partial v_i} \right) p(\boldsymbol{x}, \boldsymbol{v}, t). \tag{3.19}$$

In this section, angle brackets denote ensemble averages over the joint probability distribution $p(\boldsymbol{x}, \boldsymbol{v}, t)$ of positions and velocities at time t.

Following Ref. [18], the velocity is $v_i = \nu_i(\boldsymbol{x}) + v_i - \nu_i(\boldsymbol{x})$ for *local mean velocity*

$$\nu_i(\boldsymbol{x}) \equiv \int v_i\, p(v_i|\boldsymbol{x})\, dv_i, \tag{3.20}$$

and the entropy production rate can be decomposed into the sum of two non-negative terms:

$$\dot{\Sigma} = \sum_{i=1}^{N} D_i^{-1} \langle \nu_i(\boldsymbol{x})^2 \rangle + \sum_{i=1}^{N} D_i^{-1} \left\langle \left[v_i - \nu_i(\boldsymbol{x}) + \frac{k_{\mathrm{B}} T_i}{m_i} \frac{\partial}{\partial_{v_i}} \ln p(\boldsymbol{v}, \boldsymbol{x}, t) \right]^2 \right\rangle. \tag{3.21}$$

Here I used the fluctuation-dissipation relation (2.58) to rewrite friction coefficients in terms of diffusion coefficients.

Since both terms are non-negative, the first term is itself a lower bound for the entropy production. Applying Jensen's inequality to the first term yields subsystem and total Jensen bounds for underdamped Langevin dynamics:

$$\dot{\Sigma}_i \geq D_i^{-1} \langle v_i \rangle^2, \tag{3.22a}$$

$$\dot{\Sigma} \geq \sum_{i=1}^{N} D_i^{-1} \langle v_i \rangle^2. \tag{3.22b}$$

3.6 Non-multipartite Dynamics

The derivations thus far have relied on the assumption of multipartite dynamics, namely that the friction ($\boldsymbol{\zeta}$), mobility ($\boldsymbol{\mu}$), and diffusion ($\boldsymbol{D}$) tensors are all diagonal. I now turn to the case where the multipartite assumption breaks down such that $\boldsymbol{\mu}$, $\boldsymbol{\zeta}$, and $\boldsymbol{D}$ may have non-zero off-diagonal elements. Let us restrict our attention to the isothermal case where all fluctuations arise from coupling to heat baths at temperature T, and to the case where all diffusion, friction, and mobility coefficients are constant. Crucially, I still assume the multidimensional fluctuation-dissipation theorem

$$\boldsymbol{\zeta}^{-1} = \boldsymbol{\mu} = \boldsymbol{D}/(k_B T). \tag{3.23}$$

Without the multipartite assumption, defining the rate of entropy production becomes much less straightforward. While Ref. [19] sketches a derivation of the eventual result, Eq. (3.30), the next two subsections provide a full derivation for pedagogical purposes. Starting from the definition of total entropy production as a sum of changes in internal entropy (scaled by k_B) and heat dissipated to external baths [20],

$$\dot{\Sigma} \equiv \dot{S} - \dot{Q}/\left(k_B T\right), \tag{3.24}$$

I then proceed to evaluate the two terms $\dot{S}$ and $\dot{Q}$.

3.6.1 *Defining the Heat Flow*

Non-multipartite dynamics pose significant difficulties for identifying heat flows. In the previous sections I assumed one-to-one interactions between system components and heat baths. This is incompatible with non-multipartite dynamics, so a given system component can now be subjected to fluctuations from distinct sources. It is for this reason that to make headway I assume a single temperature T, since this allows us to ascribe all fluctuations to a single global heat bath.

To define the heat flow, start with the rate of change $\dot{E}$ of the energy, which can be decomposed into a sum of two contributions that can be identified as rates of work and heat:

$$\dot{E} = \frac{d}{dt}\langle V(\boldsymbol{x},t)\rangle \tag{3.25a}$$

$$= \int d\boldsymbol{x}\, p(\boldsymbol{x},t)\frac{\partial}{\partial t}V(\boldsymbol{x},t) + \int d\boldsymbol{x}\, V(\boldsymbol{x},t)\frac{\partial}{\partial t}p(\boldsymbol{x},t) \tag{3.25b}$$

$$= \underbrace{\int d\boldsymbol{x}\, p(\boldsymbol{x},t)\frac{\partial}{\partial t}V(\boldsymbol{x},t) + \left\langle \boldsymbol{f}_{\mathrm{nc}}(\boldsymbol{x},t)^{\top} \circ \dot{\boldsymbol{x}}\right\rangle}_{\dot{W}}$$
$$+ \underbrace{\int d\boldsymbol{x}\, V(\boldsymbol{x},t)\frac{\partial}{\partial t}p(\boldsymbol{x},t) - \left\langle \boldsymbol{f}_{\mathrm{nc}}(\boldsymbol{x},t)^{\top} \circ \dot{\boldsymbol{x}}\right\rangle}_{\dot{Q}}. \tag{3.25c}$$

Inserting the Fokker-Planck equation (2.59) and integrating the heat term by parts gives the standard definition of heat for multipartite dynamics:

$$\dot{Q} = -\left\langle \boldsymbol{f}(\boldsymbol{x},t)^{\top}\boldsymbol{J}(\boldsymbol{x},t)/p(\boldsymbol{x},t)\right\rangle. \tag{3.26}$$

3.6.2 *Defining the Entropy Production Rate*

I now compute the mean rate of change of system entropy, using the FPE (2.59) following the approach of Ref. [20], as

$$\dot{S} \equiv \frac{\mathrm{d}}{\mathrm{d}t}\left\langle -\ln p(\boldsymbol{x},t)\right\rangle \tag{3.27a}$$

$$= \left\langle -\frac{\partial_t p(\boldsymbol{x},t)}{p(\boldsymbol{x},t)} - \frac{\nabla^{\top} p(\boldsymbol{x},t)}{p(\boldsymbol{x},t)} \circ \dot{\boldsymbol{x}}\right\rangle \tag{3.27b}$$

$$= \left\langle -\frac{\nabla^{\top} p(\boldsymbol{x},t)}{p(\boldsymbol{x},t)} \circ \dot{\boldsymbol{x}}\right\rangle. \tag{3.27c}$$

In the third line the first term has been integrated out, since probability conservation imposes $\int \partial_t p(\boldsymbol{x},t)d\boldsymbol{x} = 0$.

This expression can be simplified further using the definition of the probability flux vector (2.60), rearranged to read

$$-\nabla^{\top} p(\boldsymbol{x},t) = -\frac{1}{k_{\mathrm{B}}T}\boldsymbol{f}(\boldsymbol{x},t)^{\top} p(\boldsymbol{x},t) + \left(\boldsymbol{D}^{-1}\right)^{\top}\boldsymbol{J}(\boldsymbol{x},t)^{\top}, \tag{3.28}$$

which yields

$$\dot{S} = -\frac{1}{k_{\mathrm{B}}T}\left\langle \boldsymbol{f}(\boldsymbol{x},t)^{\top} \circ \dot{\boldsymbol{x}}\right\rangle + \left\langle \frac{\boldsymbol{J}(\boldsymbol{x},t)^{\top}}{p(\boldsymbol{x},t)}\left(\boldsymbol{D}^{-1}\right)^{\top} \circ \dot{\boldsymbol{x}}\right\rangle \tag{3.29a}$$

$$= -\frac{1}{k_{\mathrm{B}}T}\left\langle \boldsymbol{f}(\boldsymbol{x},t)^{\top}\frac{\boldsymbol{J}(\boldsymbol{x},t)}{p(\boldsymbol{x},t)}\right\rangle + \left\langle \frac{\boldsymbol{J}(\boldsymbol{x},t)^{\top}\boldsymbol{D}^{-1}\boldsymbol{J}(\boldsymbol{x},t)}{p(\boldsymbol{x},t)^2}\right\rangle. \tag{3.29b}$$

In the last line I replaced the ensemble-averaged Stratonovich multiplication by $\dot{\boldsymbol{x}}$ with ensemble-averaged multiplication by the local mean velocity $\boldsymbol{J}(\boldsymbol{x},t)/p(\boldsymbol{x},t)$ [3], and replaced $\left(\boldsymbol{D}^{-1}\right)^{\top}$ with $\boldsymbol{D}^{-1}$ since the quadratic form in the second term is unchanged by taking the transpose of the matrix.

Finally, defining the total rate of entropy production as the sum of heat dissipated to the bath and system entropy change,

$$\dot{\Sigma} \equiv \dot{S} - \dot{Q}/(k_{\mathrm{B}}T) \tag{3.30a}$$

$$= \left\langle \frac{\boldsymbol{J}(\boldsymbol{x},t)^{\top}\boldsymbol{D}^{-1}\boldsymbol{J}(\boldsymbol{x},t)}{p(\boldsymbol{x},t)^2}\right\rangle. \tag{3.30b}$$

This agrees with the result derived in Ref. [19] and the result reported in Ref. [21]. Unlike in the case of multipartite dynamics discussed previously, for non-multipartite dynamics it is not generally possible to define non-negative entropy production rates at the subsystem level [22].

3.6.3 Jensen Lower Bound

While more complicated than the multipartite case, the functional form for the entropy production rate obtained in Eq. (3.30) is still amenable to lower-bounding via Jensen's inequality. For the second law ($\dot{\Sigma} \geq 0$) to hold for any flux vector $\boldsymbol{J}$, the inverse of the diffusion matrix $\boldsymbol{D}^{-1}$ (or equivalently the friction matrix $\boldsymbol{\zeta} = (\beta \boldsymbol{D})^{-1}$) must be positive semidefinite, a standard assumption for multi-dimensional friction matrices in stochastic dynamics [21, 23, 24]. Note that the expression inside the angle brackets in Eq. (3.30b) is a quadratic form in the vector $\boldsymbol{J}(\boldsymbol{x},t)/p(\boldsymbol{x},t)$ [25]. When $\boldsymbol{D}^{-1}$ is positive semidefinite, the quadratic form is a convex function of $\boldsymbol{J}(\boldsymbol{x},t)/p(\boldsymbol{x},t)$. Thus Jensen's inequality yields a general lower bound on the entropy production rate:

$$\dot{\Sigma} = \left\langle \frac{\boldsymbol{J}(\boldsymbol{x},t)^{\top}\boldsymbol{D}^{-1}\boldsymbol{J}(\boldsymbol{x},t)}{p(\boldsymbol{x},t)^2}\right\rangle \tag{3.31a}$$

$$\geq \left\langle \frac{\boldsymbol{J}(\boldsymbol{x},t)}{p(\boldsymbol{x},t)}\right\rangle^{\top} \boldsymbol{D}^{-1} \left\langle \frac{\boldsymbol{J}(\boldsymbol{x},t)}{p(\boldsymbol{x},t)}\right\rangle \tag{3.31b}$$

$$= \langle \dot{\boldsymbol{x}} \rangle^{\top} \boldsymbol{D}^{-1} \langle \dot{\boldsymbol{x}} \rangle. \tag{3.31c}$$

The result is a generalized Jensen bound depending only on the diffusion matrix and mean rates of change of the N coordinates. This reduces to Eq. (3.5b) for multipartite dynamics when the diffusion matrix (or equivalently the friction matrix) is diagonal.

When all system components have the same mean velocity v, as is the case for collective-transport systems [26, 27], Eq. (3.31c) significantly simplifies to

$$\dot{\Sigma} \geq \langle v \rangle^2 \sum_{i,j} D_{ij}^{-1}. \tag{3.32}$$

When the sum of all off-diagonal terms is non-negative, this can be further lower-bounded by the total Jensen bound for multipartite dynamics,

$$\dot{\Sigma} \geq \langle v \rangle^2 \sum_{i=1}^{N} D_{ii}^{-1}. \tag{3.33}$$

3.6.4 Alternate Derivation from the Multidimensional TUR

It is also possible to derive Eq. (3.31c) by taking a short-time limit of the multidimensional TUR. As derived in Ref. [28], the multidimensional TUR bounds the total entropy production rate of a system obeying overdamped Langevin dynamics without requiring the multipartite assumption:

$$\Sigma = \int_0^{t_f} \dot{\Sigma}\, \mathrm{d}t \geq 2 \left\langle \boldsymbol{j}_d \right\rangle^{\top} \boldsymbol{C}^{-1} \left\langle \boldsymbol{j}_d \right\rangle. \tag{3.34}$$

Here $\boldsymbol{j}_d$ is a time-integrated current of the form

$$\boldsymbol{j}_d = \int_0^{t_f} \boldsymbol{d}(\boldsymbol{x}, t)^{\top} \circ \dot{\boldsymbol{x}}\, \mathrm{d}t \tag{3.35}$$

for any function $\boldsymbol{d}(\boldsymbol{x}, t)$, and $\boldsymbol{C}$ is the covariance matrix for the current $\boldsymbol{j}_d$, defined as

$$C_{ik} = \left\langle j_{d_i} j_{d_k} \right\rangle - \left\langle j_{d_i} \right\rangle \left\langle j_{d_k} \right\rangle. \tag{3.36}$$

In the limit $t_f = \mathrm{d}t \to 0$, we recover $\Sigma = \dot{\Sigma}\, \mathrm{d}t$ and

$$\boldsymbol{j}_d \mathrm{d}_t = \boldsymbol{d}(\boldsymbol{x}, t)^{\top} \circ \mathrm{d}\boldsymbol{x}(t). \tag{3.37}$$

Setting $d_i(\boldsymbol{x}, t) = 1$ for all i yields $\left\langle \boldsymbol{j}_d \right\rangle = \langle \dot{\boldsymbol{x}} \rangle$, and $C_{ik} \mathrm{d}_t = 2 D_{ik}$. Thus we recover the Jensen bound for non-multipartite dynamics, Eq. (3.31c), so long as the diffusion coefficients are constant.

3.7 Discussion

In this chapter I derived Jensen bounds on both subsystem and total entropy production rates for multipartite overdamped Langevin dynamics (Eqs. (3.5a) and (3.5b)). I also derived extensions in several directions, allowing for position-dependent diffusion coefficients (Eqs. (3.9) and (3.10)), multipartite underdamped dynamics (Eqs. (3.22a) and (3.22b)), and non-multipartite overdamped dynamics (Eq. (3.31c)). These results make the Jensen bound applicable to a wide class of stochastic systems. I will apply the Jensen bound to study molecular machines in later chapters, using Eq. (3.5b) to derive performance bounds in Chap. 5, and Eq. (3.5a) to perform inference in Chap. 6.

> **Fundamental Limit**: The Jensen bound quantifies a minimum entropic cost for motion against viscous friction. Moreover this entropic cost comes at the subsystem level.

In all cases, the Jensen bound lower-bounds the entropy production rate of a continuous stochastic system in terms of only bare diffusion coefficients and mean rates of change of the system's degrees of freedom. These quantities can all be estimated from trajectory data: diffusion coefficients can be inferred using statistical methods [29, 30], and mean rates of change can be computed directly by taking time- or ensemble-averages. In all of the results in this chapter, the diffusion coefficients can be replaced by friction coefficients using the fluctuation-dissipation relation (2.58).

As a further application, note that the Jensen bound can be used to estimate entropy production rates in systems where they cannot be calculated exactly. Explicitly calculating the entropy production rate of a stochastic system requires knowledge of the full nonequilibrium probability distribution along with all of the conservative and non-conservative forces acting on the system. These are all in general difficult to measure or compute for systems with many degrees of freedom. The Jensen bound, however, is much more straightforward to compute from data, provided the friction/diffusion coefficients are previously known or can be measured: mean coordinate rates of change can be computed from far less data than would be required to compute the full nonequilibrium probability distribution.

> **Inference Tool**: The Jensen bound can be used to estimate subsystem and total entropy production rates from only measurements of average coordinate rates of change and friction coefficients.

It is natural to ask how the Jensen bound relates to the thermodynamic uncertainty relation (TUR), of which there are both short-time [15] and long-time [5] formulations. I showed in Sects. 3.4 and 3.6.4 that the subsystem and total Jensen bounds for overdamped Langevin dynamics can be derived from short-time TURs, meaning that the Jensen bound will generally be equal to or looser than the short-time TUR for optimal choices of current. Chapter 5 compares the Jensen bound to the long-time TUR [5] for a numerically simulated model of collective motor-driven transport, finding that there is no general hierarchy between the two bounds: either can be tighter, depending on the regime explored, with the long-time TUR tighter when current fluctuations are small, and the Jensen bound tighter otherwise. Chapter 5 also shows that the Jensen bound is always saturated for linear collective-transport systems.

For applications to experimental data, the main difference in the utilities of these different entropy production bounds arises from the different information required to compute them. The TUR and its multidimensional generalizations [28, 31–33] generally require measurements of variances and covariances of currents, but do not require detailed knowledge (e.g., friction/diffusion coefficients, conservative and nonconservative forces) regarding the equations of motion from which those dynamics arise. By contrast, the Jensen bound requires measurements of mean coordinate rates of change and knowledge of the friction/diffusion coefficients characterizing the system. While the TUR can make use of any currents in the system (including heat or energy currents), the Jensen bound only holds for coordinate rates of change. Ultimately the choice of which bounds to use should be made through considerations of what information and experimental data are available to the user.

References

1. Carnot S (1824) Reflections on the motive power of fire, and on machines fitted to develop that power. Paris Bachelier 108:1824
2. Seifert U (2019) From stochastic thermodynamics to thermodynamic inference. Annu Rev Condens Matter Phys 10:171–192
3. Seifert U (2012) Stochastic thermodynamics, fluctuation theorems and molecular machines. Rep Prog Phys 75(12):126001
4. Barato AC, Seifert U (2015) Thermodynamic uncertainty relation for biomolecular processes. Phys Rev Lett 114(15):158101
5. Gingrich TR, Horowitz JM, Perunov N, England JL (2016) Dissipation bounds all steady-state current fluctuations. Phys Rev Lett 116(12):120601
6. Horowitz JM, Gingrich TR (2020) Thermodynamic uncertainty relations constrain non-equilibrium fluctuations. Nat Phys 16(1):15–20
7. Pietzonka P, Barato AC, Seifert U (2016) Universal bound on the efficiency of molecular motors. J Stat Mech Theory Exp 12:124004
8. Nakazato M, Ito S (2021) Geometrical aspects of entropy production in stochastic thermodynamics based on Wasserstein distance. Phys Rev Res 3(4):043093
9. Shiraishi N, Funo K, Saito K (2018) Speed limit for classical stochastic processes. Phys Rev Lett 121(7):070601
10. Falasco G, Esposito M (2020) Dissipation-time uncertainty relation. Phys Rev Lett 125(12):120604
11. Van Tan V, Hasegawa Y et al (2020) Unified approach to classical speed limit and thermodynamic uncertainty relation. Phys Rev E 102(6):062132
12. Cover TM (1999) Elements of information theory. John Wiley & Sons
13. Van den Broeck C, Esposito M (2010) Three faces of the second law. II. Fokker-Planck formulation. Phys Rev E 82(1):011144
14. Mukhopadhyay N (2020) Probability and statistical inference. CRC Press

15. Otsubo S, Ito S, Dechant A, Sagawa T (2020) Estimating entropy production by machine learning of short-time fluctuating currents. Phys Rev E 101(6):062106
16. Van Tan V, Hasegawa Y (2019) Uncertainty relations for underdamped Langevin dynamics. Phys Rev E 100(3):032130
17. Dechant A, Sasa S-I (2018) Entropic bounds on currents in Langevin systems. Phys Rev E 97(6):062101
18. Dechant A (2022) Bounds on the precision of currents in underdamped Langevin dynamics. arXiv:2202.10696
19. Maes C, Netočný K, Wynants B (2008) Steady state statistics of driven diffusions. Phys A Stat Mech Appl 387(12):2675–2689
20. Seifert U (2005) Entropy production along a stochastic trajectory and an integral fluctuation theorem. Phys Rev Lett 95(4):040602
21. Ito S, Dechant A (2020) Stochastic time evolution, information geometry, and the Cramér-Rao bound. Phys Rev X 10(2):021056
22. Chetrite R, Rosinberg ML, Sagawa T, Tarjus G (2019) Information thermodynamics for interacting stochastic systems without bipartite structure. J Stat Mech Theory Exp 11:114002
23. Cichocki B, Jones RB, Kutteh R, Wajnryb E (2000) Friction and mobility for colloidal spheres in stokes flow near a boundary: the multipole method and applications. J Chem Phys 112(5):2548–2561
24. Ao P (2004) Potential in stochastic differential equations: novel construction. J Phys A Math Gen 37(3):L25
25. Boyd SP, Vandenberghe L (2004) Convex optimization. Cambridge University Press
26. Leighton MP, Sivak DA (2022) Performance scaling and trade-offs for collective motor-driven transport. New J Phys 24(1):013009
27. Leighton MP, Sivak DA (2022) Dynamic and thermodynamic bounds for collective motor-driven transport. Phys Rev Lett 129(11):118102
28. Dechant A (2018) Multidimensional thermodynamic uncertainty relations. J Phys A Math Theor 52(3):035001
29. Dechant A (2019) Estimating the free-space diffusion coefficient of trapped particles. Eur Lett 125(2):20010
30. Frishman A, Ronceray P (2020) Learning force fields from stochastic trajectories. Phys Rev X 10(2):021009
31. Dechant A, Sasa S-I (2021) Improving thermodynamic bounds using correlations. Phys Rev X 11(4):041061
32. Koyuk T, Seifert U (2022) Thermodynamic uncertainty relation in interacting many-body systems. Phys Rev Lett 129(21):210603
33. Tanogami T, Van Tan V, Saito K (2023) Universal bounds on the performance of information-thermodynamic engine. Phys Rev Res 5(4):043280

Chapter 4
Performance Scaling and Trade-Offs for Collective Motor-Driven Transport

Motor-driven intracellular transport of organelles, vesicles, and other molecular cargo is a highly collective process. An individual cargo is often pulled by a team of transport motors, with numbers ranging from only a few to several hundred. This chapter explores the behaviour of these systems using a stochastic model for transport of molecular cargo by an arbitrary number N of motors obeying linear Langevin dynamics, finding analytic solutions for the N-dependence of the velocity, precision of forward progress, energy flows between different system components, and efficiency. In two opposing regimes, I show that these properties obey simple scaling laws with N. Finally, I explore trade-offs between performance metrics as N is varied, providing insight into how different numbers of motors might be well-matched to distinct contexts where different performance metrics are prioritized.

4.1 Introduction

Motor proteins within cells often work collectively to transport large organelles such as mitochondria [1, 2], or even chromosomes during mitosis and meiosis [3]. Experimental determination of the number N of motors attached to a given cargo is generally challenging; nonetheless, recent studies have successfully measured N by identifying discrete peaks in a distinctly multimodal velocity distribution [4, 5] for small numbers of motors, or using more complex techniques such as quantitative immunoblots and immunoelectron microscopy [2]. Experimental investigations both *in vivo* and *in vitro* have found widely varying numbers N of motors coupled to a single cargo. In some cases, only a single motor [6] or a few motors [4] per cargo is observed, but experiments have observed as many as 200 motors bound to large organelles [2]. Likewise, in actomyosin filaments in muscle tissue, on the order of 100 motors are attached to each actin filament [7]. Collective-transport systems can

M. Leighton, *Stochastic Thermodynamics of Multicomponent Molecular Machines*, Springer Theses, https://doi.org/10.1007/978-3-032-02204-2_4

also be engineered *in vitro* [8]; in this setting the number of motors can be controlled more precisely, for example using DNA scaffolds [9, 10].

Simple phenomenological models for transport-motor dynamics (such as the classical linear force-velocity relationship [6, 11]) have been extended to multiple motors, for example by assuming equal load-sharing [4, 5]. These models provide a good first approximation to the dynamics of multi-motor systems, and can be extended to include, for example, motor binding and unbinding kinetics [12, 13]. These types of models assume that the motors pull against a constant force, rather than the stochastic frictional drag that would occur for a loosely coupled diffusive cargo. However, analysis of transport by single motors has shown that pulling a diffusing cargo and pulling against a constant force lead to qualitatively different transport behaviour [14, 15]. Researchers have proposed and explored several dynamical models for transport of diffusive cargo by multiple motors [16–21]. These approaches rely primarily on numerical simulation, and as such are limited by computational resources to exploring systems with relatively small numbers of motors.

In studying intracellular transport, an important goal is to understand how systems can be tuned to improve performance. Relevant performance metrics vary based on the context, but may include cargo velocity, rate of chemical energy consumption, transport efficiency, and precision [22]. The dependence of these and other performance metrics on the number of motors is of clear interest, and has not yet been systematically investigated.

In this chapter, I introduce a simple, thermodynamically consistent, stochastic model for the collective transport of diffusive molecular cargo by an arbitrary number N of motors. This model has a key advantage over other recent theoretical and computational approaches: it is analytically tractable for arbitrary N, allowing us to explore system behaviour over many orders of magnitude of motor numbers. I derive N-dependent expressions for several performance metrics, and explicitly calculate all thermodynamic energy flows between different system components and thermal and chemical reservoirs. This allows us to derive simple analytic expressions for efficiency both of the whole system and of individual motors. In two opposing regimes I identify simple scaling laws that characterize the N-dependence of these properties. Finally, I derive fundamental trade-offs among these performance metrics, thereby pointing to design principles for collective motor-driven transport.

4.2 Model and Theory

Consider a diffusive cargo coupled to N identical transport motors, with motion resolved in one dimension. Each motor interacts only with the cargo via a molecular linker, and is characterized by a mechanochemical cycle through which it transduces chemical power into directed forward motion. The cargo undergoes Brownian motion subject to coupling forces from each motor via the respective linker. Figure 4.1 illustrates the system.

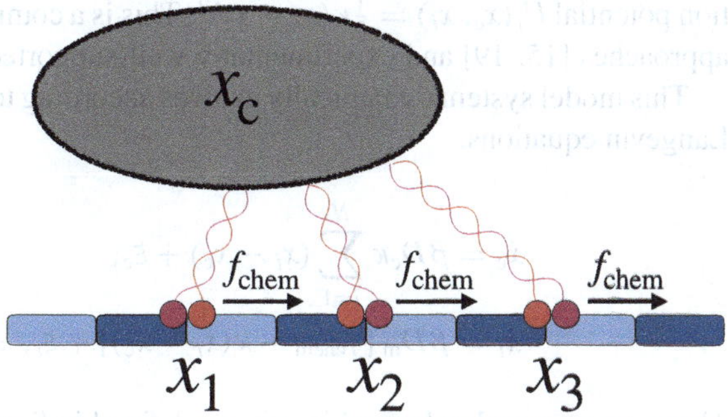

Fig. 4.1 Collective-transport system comprising a single diffusive cargo coupled to N (here $N = 3$) motor proteins moving along a linear substrate. x_c indicates the cargo position, and x_i the position of the ith motor. Each motor experiences chemical driving force f_{chem}

This chapter focuses on the limiting regime in which the time evolution of the system is independent of the initial conditions, its steady state. Mathematically, the relevant limit is that the time greatly exceeds the system's longest relaxation time ($t \gg \tau_{relax}$). I assume that this limit exists, and that in the steady state, system properties such as the velocity, efficiency, energy flows and entropy production all have well-defined constant average values. The specific model introduced below satisfies these assumptions.

4.2.1 Model

The cargo and motors are modelled as overdamped Brownian particles diffusing (with respective diffusivities D_c and D_m) in a potential landscape. x_c and x_i describe the positions of the cargo and ith motor. For mathematical simplicity the cargo is treated as a single point, but the model describes equally well (through a linear change of variables) motors attached at different points to a rigid cargo. The system is isothermal and in contact with a thermal reservoir at inverse temperature $\beta \equiv (k_B T)^{-1}$. I assume that both cargo and motor dynamics satisfy the fluctuation-dissipation relation: the friction coefficients for the cargo (ζ_c) and motors (ζ_m) are related to their respective diffusivities by $\beta\zeta_c D_c = 1 = \beta\zeta_m D_m$.

Each motor is subject to chemical driving due to nonequilibrium environmental concentrations of molecules, often ATP, ADP, and phosphate, and further experiences an underlying periodic energy landscape due to interactions with the substrate it walks along. To simplify the analysis, I assume that the scale of this energy landscape (the heights of the barriers separating meta-stable states) are small compared to the magnitude of the chemical driving. This leads to the motors experiencing a flat, downward-sloping energy landscape, which can be thought of as a constant force f_{chem} propelling each motor in a preferred direction. I further assume tight coupling between chemical-energy consumption and forward motor motion.

Each linker coupling one motor to the cargo is modelled as a Hookean spring with zero rest length, dominated by the along-filament displacement, thus with interaction potential $U_i(x_c, x_i) = \frac{1}{2}\kappa(x_i - x_c)^2$. This is a common assumption in modelling approaches [15, 19] and experimentally well-supported for kinesin linkers [23].

This model system dynamically evolves according to $N + 1$ coupled overdamped Langevin equations,

$$\dot{x}_c = \beta D_c \kappa \sum_{i=1}^{N} (x_i - x_c) + \xi_c, \tag{4.1a}$$

$$\dot{x}_i = \beta D_m [f_{chem} - \kappa(x_i - x_c)] + \xi_i, \; i = 1, \ldots, N. \tag{4.1b}$$

These are assumed to be multipartite as defined in Sec. 2.5.

The single-motor dynamics (4.1b) produce average motion equivalent to the linear force-velocity relation typically observed experimentally for kinesin motors under constant forces less than their stall force [6, 11] (where most of the analysis takes place),

$$\langle v \rangle = v_{max} \left(1 - \frac{f}{f_s}\right), \tag{4.2}$$

for stall force $f_s = f_{chem}$, maximum velocity $v_{max} = \beta D_m f_{chem}$, and force $f = \kappa \langle x_i - x_c \rangle$ acting on the motor.

4.2.2 Parameter Estimates

Experimentally, kinesin linkers are well-approximated as Hookean springs with a zero rest length and a spring constant ~0.5 pN/nm [23]. Similar behaviour has been observed for the linkers of myosin V motors, which have an estimated spring constant of $0.2 - 0.4$ pN/nm [24].

The chemical driving force f_{chem} can be estimated in two ways. By noting the equivalence in Eq. (4.2) of f_{chem} with the single-motor stall force, experimental estimates of single-motor stall forces can be used to estimate the chemical driving force. Kinesin motors have stall forces on the order of $5 - 8$ pN [25], while myosin motors stall at forces as high as 15 pN [26].

Likewise, due to tight coupling between chemical energy consumption and mechanical motion [25, 27], f_{chem} can also be thought of as a free-energy dissipation per unit distance. Kinesin, for example, hydrolyzes one molecule of ATP (a reaction with a free-energy change $\Delta\mu_{ATP} \approx 15 - 30\, k_B T$ [28]) for every forward step ($d \approx 8$ nm). At 298 K, $1\, k_B T = 4.114$ pN nm, resulting in a chemical driving force $f_{chem} = \Delta\mu_{ATP}/d \approx 8 - 15$ pN, in line with the previous estimate.

I estimate the motor diffusivity using $v_{max} = \beta f_s D_m$. For kinesin-1 the maximum velocity is $v_{max} \approx 1 - 2\, \mu$m/s and the stall force is $f_s \approx 6 - 8$ pN [29], while for myosin V, $v_{max} \approx 8$ nm/s [30] and $f_s \approx 10 - 15$ pN [7]. This suggests that in both cases $D_m = \mathcal{O}(10^{-3})\, \mu m^2/s$. Alternatively, using experimental estimates of rate constants for forward and reverse steps yields an estimate for kinesin-1 motor diffusivity of $D_{\mathrm{m}} \approx 4 \times 10^{-3}\ \mu\mathrm{m}^2/\mathrm{s}$ [31].

Cargo diffusivity can vary by orders of magnitude depending on the type of molecular cargo. As one example, diffusivity of vesicles (with radii on the order of 300 nm) in neurons is estimated to be of order $10^{-3}\ \mu\mathrm{m}^2/\mathrm{s}$ [32]. Other measurements of vesicles and vesicle-sized beads in cytoplasm have found diffusivities on the order of $10^{-4} - 10^{-2}\ \mu\mathrm{m}^2/\mathrm{s}$ [33]. Larger cargo such as organelles, for example mitochondria which have diameters as large as $2\mu\mathrm{m}$, will have even smaller diffusivities. Thus for intracellular transport of vesicles and organelles we expect $D_{\mathrm{c}}/D_{\mathrm{m}} \in [10^{-3}, 1]$.

4.3 Results

4.3.1 Solution

Equation (4.1) constitute a linear system of coupled Langevin equations, and as such are in general analytically solvable, with solution a multivariate Gaussian. Thus it suffices to solve for the mean vector and covariance matrix of the whole system, the components of which satisfy a set of coupled linear ordinary differential equations [34, Sect. 3.2]. Since the N motors dynamically evolve according to identical stochastic equations (4.1b), their marginal position distributions are identical. As a result, there are only two unique means ($\langle x_i \rangle$ and $\langle x_{\mathrm{c}} \rangle$) and four unique covariances ($\mathrm{Cov}(x_{\mathrm{c}}, x_{\mathrm{c}})$, $\mathrm{Cov}(x_{\mathrm{c}}, x_i)$, $\mathrm{Cov}(x_i, x_i)$, and $\mathrm{Cov}(x_i, x_j)$, all of which vary with time. This symmetry permits exact solution for arbitrary N. Starting from initial conditions $x_{\mathrm{c}} = x_i = 0$ at $t = 0$, at steady state ($t \gg \tau_{\mathrm{relax}}$) the mean cargo position and motor positions are

$$\langle x_{\mathrm{c}} \rangle = \langle v \rangle t - \frac{N f_{\mathrm{chem}}}{\kappa} \frac{D_{\mathrm{eff}}^2}{D_{\mathrm{m}} D_{\mathrm{c}}}, \tag{4.3a}$$

$$\langle x_i \rangle = \langle v \rangle t + \frac{f_{\mathrm{chem}}}{\kappa} \left(\frac{D_{\mathrm{eff}}}{D_{\mathrm{c}}} \right)^2, \tag{4.3b}$$

and the covariances are

$$\mathrm{Cov}(x_{\mathrm{c}}, x_{\mathrm{c}}) = 2 D_{\mathrm{eff}} t + \frac{N}{\beta\kappa} \left(\frac{D_{\mathrm{eff}}}{D_{\mathrm{m}}} \right)^2, \tag{4.4a}$$

$$\mathrm{Cov}(x_{\mathrm{c}}, x_i) = 2 D_{\mathrm{eff}} t - \frac{1}{\beta\kappa} \frac{D_{\mathrm{eff}}^2}{D_{\mathrm{m}} D_{\mathrm{c}}}, \tag{4.4b}$$

$$\mathrm{Cov}(x_i, x_j) = 2 D_{\mathrm{eff}} t + \frac{1}{\beta\kappa} \left[\delta_{ij} - \frac{D_{\mathrm{eff}}^2}{D_{\mathrm{m}}} \left(\frac{N}{D_{\mathrm{m}}} + \frac{2}{D_{\mathrm{c}}} \right) \right], \tag{4.4c}$$

for Kronecker delta function δ_{ij}. Here $\langle v \rangle$ and D_{eff} are the mean velocity and effective diffusion coefficient, which will be defined and evaluated in the following section.

Note that different initial conditions would produce different time-independent constant terms in (4.3) and (4.4); however for large times (in the steady-state limit) the constant terms are negligible compared to the terms linear in t. Regardless of initial conditions, the difference between the constant terms in (4.3)a and b will always be the mean value of the separation distance, $\langle \Delta x_i \rangle$.

The distributions of $\{x_i(t)\}_{i=1}^N$ and $x_c(t)$ are time-dependent, so we change to a set of N variables, $\Delta x_i(t) = x_i(t) - x_c(t)$, that at steady state converge to a time-independent joint distribution, a multivariate Gaussian with means and covariances

$$\langle \Delta x_i \rangle = \frac{f_{\text{chem}}}{\kappa}\left(1 + N\frac{D_c}{D_m}\right)^{-1}, \tag{4.5a}$$

$$\text{Cov}(\Delta x_i, \Delta x_j) = \frac{1}{\beta\kappa}\delta_{ij}. \tag{4.5b}$$

This time-independent distribution is sufficient to compute many steady-state properties of interest.

The off-diagonal entries of the stationary covariance matrix, $\text{Cov}(\Delta x_i, \Delta x_j)$ for $i \neq j$, are zero: fluctuations in the relative position of one motor are uncorrelated with the relative positions of the other motors. I do not expect this particular result to generalize; for example, collective-transport models with discrete motor motion [19] have found non-zero off-diagonal covariances at small N. Regardless, the results presented below depend only on diagonal terms and are independent of off-diagonal covariances.

The system relaxation time is $\tau_{\text{relax}} = [\beta\kappa(D_m + ND_c)]^{-1}$. The earlier parameter ranges and estimates yield a maximum value of about 0.02 s (taking $\kappa = 0.2$ pN/nm, $T = 298$ K, $D_m = 10^{-3}\ \mu\text{m}^2/\text{s}$, $D_c = 10^{-4}\ \mu\text{m}^2/\text{s}$, and $N = 1$). Using more typical values of these parameters (for example $\kappa = 0.5$ pN/nm, $D_m = 4 \times 10^{-3}\ \mu\text{m}^2/\text{s}$, $D_c = 10^{-3}\ \mu\text{m}^2/\text{s}$, and $N = 10$) gives a much smaller estimate $\tau_{\text{relax}} \approx 5 \times 10^{-4}$ s. A relaxation time of $\tau_{\text{relax}} = 0.02$ s corresponds to a distance of at most 40 nm for kinesin motors at maximum velocity. Given the short distance over which relaxation to steady state occurs, and the high processivity of motor-driven transport systems (kinesin can travel up to several micrometers before detaching [35]), this chapter focuses exclusively on the steady state.

The dimensionless parameter combination ND_c/D_m appears in Eq. (4.5a) and in many of the results shown later on, constituting a key quantity for understanding the system behaviour. For intracellular transport of vesicles or organelles the diffusivity ratio D_c/D_m is typically $\sim 10^{-3} - 10^0$, depending on the size of the cargo. Since N can range from one to several hundred, the explorations in this chapter focus on the range $ND_c/D_m \in [10^{-3}, 10^3]$.

4.3.2 Scaling Behaviour

Dynamical Properties

At steady state, the cargo and each motor have equal velocity, $\langle v_c \rangle \equiv \lim_{t\to\infty}\langle x_c(t) - x_c(0)\rangle/t = \lim_{t\to\infty}\langle x_i(t) - x_i(0)\rangle/t \equiv \langle v_m \rangle \equiv \langle v \rangle$. Evaluating this limit yields a simple expression for the N-dependent mean system velocity:

$$\langle v \rangle = v_{\max}\left(1 + \frac{D_m}{ND_c}\right)^{-1} . \tag{4.6}$$

For $N \ll D_m/D_c$, the mean velocity grows linearly with N, so adding more motors proportionally increases the system velocity. As N grows much larger than D_m/D_c, however, the steady-state velocity asymptotically approaches maximum velocity $v_{\max} = \beta D_m f_{\text{chem}}$ (the mean velocity of an uncoupled motor) as $\langle v \rangle \approx v_{\max}\,[1 - D_m/(ND_c)]$. Thus no matter how many motors are coupled to the cargo, the mean velocity of the aggregate motor-cargo system never exceeds that of an unladen motor. This mean velocity (as well as the maximum velocity) scales linearly with the chemical driving force f_{chem}. Figure 4.2a shows normalized velocity $\langle v \rangle / v_{\max}$ as a function of N.

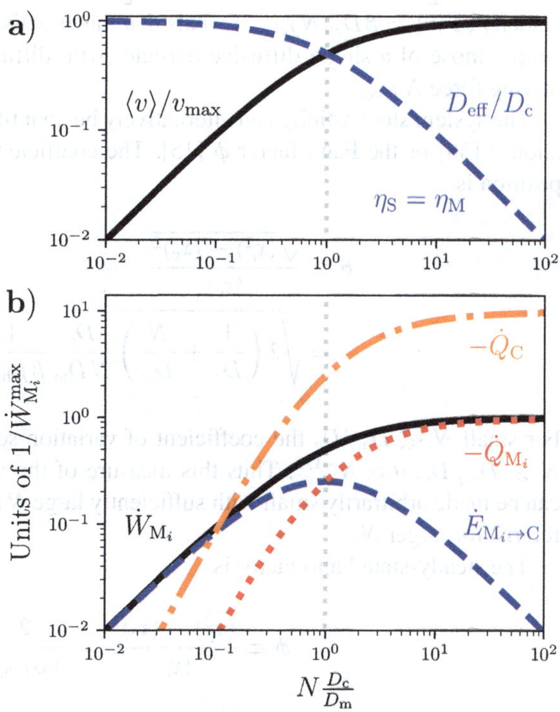

Fig. 4.2 **a** Scaled velocity $\langle v \rangle / v_{\max}$ (Eq. (4.6)) and effective diffusivity D_{eff}/D_c (Eq. (4.7)) (equal for this model to the efficiency $\eta \equiv \eta_M = \eta_S$ (Eq. (4.14))) as functions of the number N of motors scaled by the diffusivity ratio D_m/D_c. **b** Ensemble-averaged energy flows $\dot{W}_{M_i}$ (Eq. (4.10)), $\dot{E}_{M_i \to C}$ (Eq. (4.11)), $\dot{Q}_{M_i}$ (Eq. (4.12)), and $\dot{Q}_C$ (Eq. (4.13)) as functions of N scaled by D_c/D_m. All energy flows are normalized by the chemical power consumption $\dot{W}^{\max}_{M_i} = f_{\text{chem}}\, v_{\max}$ of a single motor at maximum velocity. For $\dot{Q}_C$ I use $D_c/D_m = 1/10$; all other energy flows depend solely on the quantity ND_c/D_m

While $\langle v \rangle$ gives the average motion, the model (like all transport at the cellular level) is inherently stochastic. As such, the average velocity is not sufficient to fully describe system behaviour, even at steady state. The effective cargo diffusivity D_{eff} quantifies the rate at which the variance of forward progress grows at steady state:

$$D_{\text{eff}} \equiv \lim_{t\to\infty} \frac{\langle x_c^2 \rangle - \langle x_c \rangle^2}{2t} \tag{4.7a}$$

$$= \left(\frac{1}{D_c} + \frac{N}{D_m} \right)^{-1} . \tag{4.7b}$$

Here Eq. (4.7b) is the result obtained for this model. This expression can be understood by noting that the effective friction coefficient for the system, $\zeta_{\text{eff}} = 1/\beta D_{\text{eff}}$, is simply the sum of the friction coefficients for the motors and cargo. This interpretation is consistent with a recent theoretical study of collective transport of a diffusive cargo using discrete motor dynamics, which similarly found that the contribution from the motors to the effective friction coefficient of the system scales linearly with the number N of motors for the small range explored [36]. Previous work using a simpler phenomenological model also suggested that the effective friction coefficient of a collection of motors should be proportional to N [37].

Figure 4.2a shows D_{eff} as a function of N. While the velocity increases with the number of motors, the effective diffusivity decreases, indicating that a larger number of motors tightens the distribution of cargo-transport distances. Writing the mean velocity as $\langle v \rangle = \beta D_{\text{eff}} N f_{\text{chem}}$ reveals that the effective dynamics of the system are simply those of a single diffusive particle (with diffusivity D_{eff}) under a constant driving force $N f_{\text{chem}}$.

The system stochasticity can alternatively be quantified by the coefficient of variation θ [38] or the Fano factor ϕ [15]. The coefficient of variation (CV) of cargo position is

$$\theta \equiv \frac{\sqrt{\langle x_c^2 \rangle - \langle x_c \rangle^2}}{\langle x_c \rangle} \tag{4.8a}$$

$$= \sqrt{2\left(\frac{1}{D_c} + \frac{N}{D_m} \right) \frac{D_c}{N D_m} \frac{1}{\beta f_{\text{chem}}}}\, t^{-1/2} . \tag{4.8b}$$

For small $N \ll D_m/D_c$ the coefficient of variation scales as N^{-1}, while for large $N \gg D_m/D_c$, $\theta \propto N^{-1/2}$. Thus this measure of the variation in forward progress can be made arbitrarily small with sufficiently large $N D_c/D_m$, but with diminishing returns for larger N.

The steady-state Fano factor is

$$\phi \equiv \frac{\langle x_c^2 \rangle - \langle x_c \rangle^2}{\langle x_c \rangle} = \frac{2}{N\beta f_{\text{chem}}} . \tag{4.9}$$

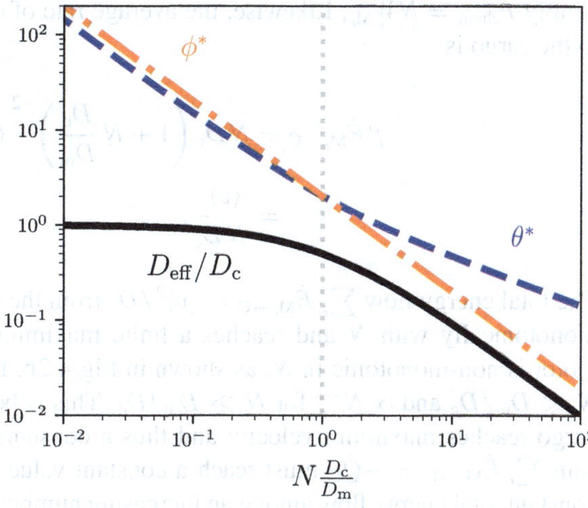

Fig. 4.3 N-dependence of the performance metrics used to quantify stochasticity: the effective diffusivity $D_{\text{eff}}/D_{\text{c}}$, the coefficient of variation $\theta^* = \theta\beta f_{\text{chem}}\sqrt{t/D_{\text{c}}}$, and the Fano factor $\phi^* = \beta f_{\text{chem}}(D_{\text{m}}/D_{\text{c}})\phi$; all in dimensionless units

Similarly to D_{eff}, ϕ decreases (and hence the precision increases) with the number of motors, scaling as $\phi \propto N^{-1}$. Here adding motors decreases the variance of forward progress while increasing the velocity, leading to a Fano factor that decreases with N. For $N = 1$ motor, Eq. (4.9) recovers the Fano factor previously calculated in the limit of low cargo diffusivity for a single motor [15].

The addition of motors can be thought of as having an "averaging" effect on the dynamics. The precision (as quantified by θ or ϕ) also increases with the chemical driving force f_{chem} on each motor. Figure 4.3 shows the N-dependence of the three stochasticity metrics: the effective diffusivity D_{eff} (Eq. (4.7)), the coefficient of variation θ (Eq. (4.8)), and the Fano factor ϕ (Eq. (4.9)).

Thermodynamic Properties

It is also possible to exactly calculate steady-state ensemble averages of all energy flows into and out of each subsystem. Transport systems are highly processive, so we focus on average energy flows, ignoring higher moments that are less salient at long durations. The mean chemical power to each motor is

$$\beta \dot{W}_{\text{M}_i} = ND_{\text{c}}\left(1 + N\frac{D_{\text{c}}}{D_{\text{m}}}\right)^{-1}(\beta f_{\text{chem}})^2 \tag{4.10a}$$

$$= \beta f_{\text{chem}}\langle v\rangle. \tag{4.10b}$$

Here the energy flows are multiplied by the inverse temperature β so that quantities have units of s^{-1}. In keeping with the assumption of tight mechanochemical coupling, the power consumption is simply the chemical driving force multiplied by the motor velocity. Since the motors are identical, the total chemical power consumption is simply $P_{\text{chem}} = N\dot{W}_{\text{M}_i}$. Likewise, the average rate of energy flow from each motor to the cargo is

$$\beta \dot{E}_{\text{M}_i \to \text{C}} = N D_{\text{c}} \left(1 + N \frac{D_{\text{c}}}{D_{\text{m}}}\right)^{-2} (\beta f_{\text{chem}})^2 \tag{4.11a}$$

$$= \frac{\langle v \rangle^2}{N D_{\text{c}}}. \tag{4.11b}$$

The total energy flow $\sum_i \dot{E}_{\text{M}_i \to \text{C}} = \langle v \rangle^2 / D_{\text{c}}$ from the motors to the cargo increases monotonically with N and reaches a finite maximum value. $\dot{E}_{\text{M}_i \to \text{C}}$ for a single motor is non-monotonic in N, as shown in Fig. 4.2b. In particular, $\dot{E}_{\text{M}_i \to \text{C}} \propto N$ for $N \ll D_{\text{m}}/D_{\text{c}}$ and $\propto N^{-1}$ for $N \gg D_{\text{m}}/D_{\text{c}}$. This is because for large N, when the cargo reaches maximum velocity and thus a constant rate of heat dissipation, the sum $\sum_i \dot{E}_{\text{M}_i \to \text{C}} = -\dot{Q}_{\text{C}}$ must reach a constant value as well. Dividing this nearly constant total energy flow among an increasing number of motors means that $\dot{E}_{\text{M}_i \to \text{C}}$ decreases. Thus the energy flow from each individual motor to the cargo is maximized at an intermediate $N^* = D_{\text{m}}/D_{\text{c}}$.

The average heat flow into each motor is

$$\beta \dot{Q}_{\text{M}_i} = -N^2 \frac{D_{\text{c}}^2}{D_{\text{m}}} \left(1 + N \frac{D_{\text{c}}}{D_{\text{m}}}\right)^{-2} (\beta f_{\text{chem}})^2 \tag{4.12a}$$

$$= -\frac{\langle v \rangle^2}{D_{\text{m}}}, \tag{4.12b}$$

and the heat flow into the cargo is

$$\beta \dot{Q}_{\text{C}} = -N^2 D_{\text{c}} \left(1 + N \frac{D_{\text{c}}}{D_{\text{m}}}\right)^{-2} (\beta f_{\text{chem}})^2 \tag{4.13a}$$

$$= -\frac{\langle v \rangle^2}{D_{\text{c}}}. \tag{4.13b}$$

The subsystem-specific heat flows are, on average, given by the friction coefficient (e.g. $\zeta_{\text{c}} = 1/\beta D_{\text{c}}$ for the cargo) multiplied by the mean velocity squared, the result we would expect for the frictional energy dissipation of an overdamped particle moving at constant velocity $\langle v \rangle$. The sum of the heat flows over all $N + 1$ subsystems represents the total energy dissipation of the system at steady state; all of the chemical energy consumed by the motors is either dissipated directly by the motors as heat (due to loose coupling to the cargo) or transduced into mechanical work on the cargo which is then dissipated by the cargo as heat.

Figure 4.2b shows how these four steady-state energy flows depend on the number N of motors, manifesting two regimes. For $N \ll D_{\text{m}}/D_{\text{c}}$, the heats $-\dot{Q}_{\text{M}_i}$ and $-\dot{Q}_{\text{C}}$ scale as N^2, while the chemical power $\dot{W}_{\text{M}_i}$ to each motor and the power $\dot{E}_{\text{M}_i \to \text{C}}$ from each motor to the cargo scale linearly with N. For $N \gg D_{\text{m}}/D_{\text{c}}$, the chemical power to each motor as well as the two heats asymptotically approach constants. For sufficiently large N, each motor's heat roughly equals its chemical power consumption, as the energy flow per motor transferred through the coupling decays as N^{-1}.

All energy flows display the same quadratic dependence on the chemical driving force. This is reminiscent of linear irreversible thermodynamics, where rates of entropy production (and thus heat dissipation) are quadratic in the thermodynamic driving forces [39]. This is true for this system on average due to the linearity of Eqs. (4.1)a and b, even with the inherent stochasticity.

The steady-state energy flows (4.10)–(4.13) are all independent of the coupling strength κ, despite the steady-state distributions for the separation distances Δx_i depending strongly on κ. To understand this initially surprising result, consider the average force an individual motor pulls against, $\kappa\langle\Delta x_i\rangle$, for separation distance $\Delta x_i \equiv x_i - x_c$. Equation (4.5a) shows that $\langle\Delta x_i\rangle \propto 1/\kappa$, so the mean force on the motor is independent of the coupling strength. Since the motor velocity is also independent of the coupling strength, the mean power output of each motor (roughly the mean velocity multiplied by the mean opposing force) is independent as well. The power consumption $P_{\to M_i}$ is likewise independent of κ for the same reason. From the first laws (2.62), the system only has three independent energy flows, so the other energy flows must thus also be independent of κ.

I consider several different metrics of steady-state energetic efficiency. Since the system does not perform any thermodynamic work as output, and the input power P_{chem} is always positive, the full system's thermodynamic efficiency is zero; however, the thermodynamic efficiency $\eta_{\text{M}} \equiv \dot{E}_{\text{M}_i\to\text{C}}/W_{\text{M}_i}$ of each motor subsystem is positive. The Stokes efficiency, given by Eq. (5.3), is also positive. This system has equal Stokes efficiency and motor efficiency:

$$\eta_{\text{M}} = \eta_{\text{S}} = \left(1 + N\frac{D_{\text{c}}}{D_{\text{m}}}\right)^{-1}. \tag{4.14}$$

Figure 4.2a shows efficiency as a function of N: for small $ND_{\text{c}}/D_{\text{m}}$, the efficiency is $\approx 1 - ND_{\text{c}}/D_{\text{m}}$, while for $ND_{\text{c}}/D_{\text{m}} \gg 1$ the efficiency scales as N^{-1}. For a given diffusivity ratio $D_{\text{c}}/D_{\text{m}}$, the efficiency for any N is upper bounded by $\eta_{\text{max}} = (1 + D_{\text{c}}/D_{\text{m}})^{-1}$ since N is lower bounded by unity. Thus, for example, a system with $D_{\text{c}} = D_{\text{m}}$ can achieve at most 50% efficiency. This efficiency can be re-written in terms of the system's effective diffusivity as

$$\eta_{\text{M}} = \eta_{\text{S}} = \frac{D_{\text{eff}}}{D_{\text{c}}}, \tag{4.15}$$

which exactly saturates an upper bound proven for the Stokes efficiency of transport by a single motor [40].

Table 4.1 summarizes the scaling with N of key performance metrics in the two limiting regimes.

Table 4.1 Performance metrics' asymptotic scaling with N

Metric	$N \ll D_m/D_c$	$N \gg D_m/D_c$
$\langle v \rangle$	$\propto N$	$\approx v_{max}\,[1 - D_m/(N D_c)]$
θ	$\propto N^{-1}$	$\propto N^{-1/2}$
$\dot{W}_{chem}$	$\propto N^2$	$\propto N$
$\eta_{S/M}$	$\approx 1 - N D_c/D_m$	$\propto N^{-1}$

4.3.3 Performance Trade-Offs

The previous section outlined the separate N-dependence of different performance metrics; however, these quantities are not independent, instead posing trade-offs parameterized by N. Examining the trade-offs between all pairs of dynamical and thermodynamic properties from Table 4.1 shows that several pairs of desirable properties cannot be simultaneously attained.

The mean transport velocity and the total chemical power consumption of the motors are related by

$$\frac{\dot{W}_{chem}}{\dot{W}^{max}_{M_i}} = \frac{D_m}{D_c} \frac{(\langle v \rangle / v_{max})^2}{1 - \langle v \rangle / v_{max}}. \tag{4.16}$$

Figure 4.4a illustrates this trade-off as N is varied, for several different diffusivity ratios. For $N \ll D_m/D_c$, the total chemical input power scales as the square of the average velocity $\langle v \rangle$. At $N = D_m/D_c$ the velocity is half its maximum; beyond this velocity the required chemical power skyrockets, scaling as $\dot{W}_{chem} \propto (v_{max} - \langle v \rangle)^{-1}$ for $N \gg D_m/D_c$.

The total power consumption and coefficient of variation (4.8) are inversely related,

$$\dot{W}_{chem} = (2/t)\theta^{-2}, \tag{4.17}$$

for all N and D_c/D_m. Figure 4.4b illustrates this trade-off, which features scaling behaviour consistent across the regimes of large and small N. Arbitrarily high precision (low θ) can be achieved in this collective-transport system, but at the cost of power consumption that increases without bound.

Comparing (4.6) and (4.14), the efficiency and scaled velocity obey a simple relation:

$$\eta_{S/M} + \frac{\langle v \rangle}{v_{max}} = 1, \tag{4.18}$$

where $\eta_{S/M}$ can be either the Stokes or motor efficiency, since they are equal for this system. Figure 4.4c shows this trade-off. The collective-transport system cannot simultaneously achieve high efficiency and near-maximum velocity. Further, depending on the diffusivity ratio only certain efficiencies are achievable: those with

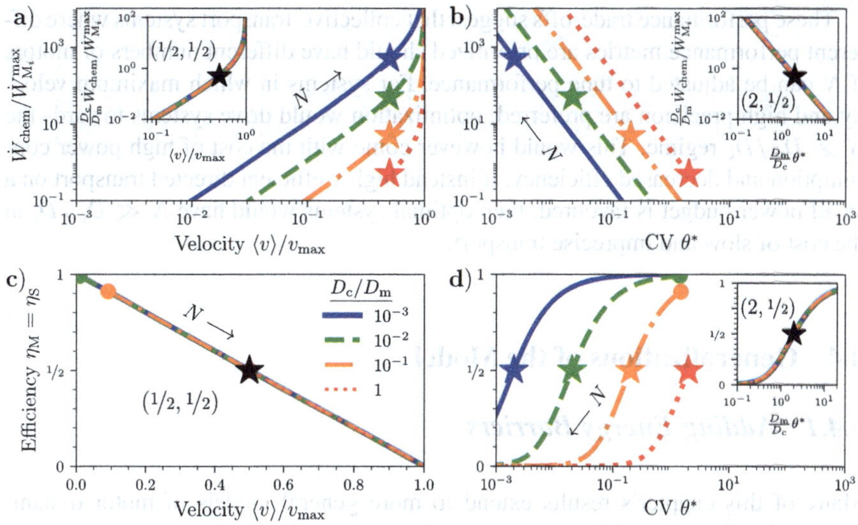

Fig. 4.4 Trade-offs between **a** scaled chemical power consumption $\dot{W}_{\mathrm{chem}}/\dot{W}^{\mathrm{max}}_{\mathrm{M}_i}$ and scaled mean velocity $\langle v\rangle/v_{\mathrm{max}}$, **b** $\dot{W}_{\mathrm{chem}}/\dot{W}^{\mathrm{max}}_{\mathrm{M}_i}$ and scaled coefficient of variation $\theta^* = \theta v_{\mathrm{max}}\sqrt{t/D_{\mathrm{c}}}$, **c** efficiency $\eta_{\mathrm{S}} = \eta_{\mathrm{M}}$ and $\langle v\rangle/v_{\mathrm{max}}$, and **d** $\eta_{\mathrm{S}} = \eta_{\mathrm{M}}$ and θ^*, as the number N of motors is varied, for different diffusivity ratios $D_{\mathrm{c}}/D_{\mathrm{m}}$. Stars: $N = D_{\mathrm{m}}/D_{\mathrm{c}}$. Circles: maximum efficiencies for respective diffusivity ratios, $\eta_{\mathrm{max}} = (1 + D_{\mathrm{c}}/D_{\mathrm{m}})^{-1}$, realized for $N = 1$. Insets in (**a**), (**b**), and (**d**) show that scaling the power consumption and coefficient of variation by factors of $D_{\mathrm{c}}/D_{\mathrm{m}}$ collapses the curves for different diffusivity ratios onto single master curves. Numbers in parentheses indicate coordinates of black stars (where $N D_{\mathrm{c}}/D_{\mathrm{m}} = 1$). All curves terminate at points where $N = 1$, so there is always an integer number of motors

$\eta_{\mathrm{S/M}} \leq \eta_{\mathrm{max}} = (1 + D_{\mathrm{c}}/D_{\mathrm{m}})^{-1}$. Notably, 50% efficiency and half-maximal velocity can always be achieved at $N = D_{\mathrm{m}}/D_{\mathrm{c}}$.

Finally consider the trade-off between efficiency and precision (quantified by coefficient of variation θ),

$$\theta\, v_{\mathrm{max}}\sqrt{\frac{t}{2D_{\mathrm{c}}}} = \frac{\sqrt{\eta_{\mathrm{S/M}}}}{1-\eta_{\mathrm{S/M}}}. \tag{4.19}$$

Figure 4.4d shows that high efficiency and high precision (low CV) are not simultaneously achievable. This suggests that to maximize efficiency systems must exploit thermal fluctuations, leading to a decrease in precision. Note that the transition from near-zero to near-unit efficiency occurs over a small range of CVs around $N = D_{\mathrm{m}}/D_{\mathrm{c}}$.

Insets in Fig. 4.4a, b, and d show that scaling the power consumption and coefficient of variation by factors of the diffusivity ratio $D_{\mathrm{c}}/D_{\mathrm{m}}$ collapses the separate curves for distinct $D_{\mathrm{c}}/D_{\mathrm{m}}$ onto single master curves. Thus the qualitative nature of the trade-offs described here are independent of the relative diffusivities of the motors and cargo.

These performance trade-offs suggest that collective-transport systems where different performance metrics are prioritized should have different numbers of motors if N can be adjusted to tune performance. For systems in which maximum velocity and high precision are preferred, optimization would drive systems towards the $N \gg D_{\rm m}/D_{\rm c}$ regime. This would however come with the cost of high power consumption and decreased efficiency. If instead highly efficient directed transport on a small power budget is favoured, then optimal systems would have $N \ll D_{\rm m}/D_{\rm c}$ at the cost of slow and imprecise transport.

4.4 Generalizations of the Model

4.4.1 Adding Energy Barriers

Many of this chapter's results extend to more general models of motor dynamics. To show this, consider a more complex model which relaxes the assumption that the chemical driving force is much larger than the scale of the motor's energy landscape. We add to each motor a periodic potential-energy landscape of the form $V(x_i) = (1/2)E^{\ddagger} \cos(2\pi x_i/\ell)$, where $E^{\ddagger}$ is the height of the energy barriers between successive meta-stable states (local energy minima), and ℓ is the period. Each motor's dynamics satisfy

$$\dot{x}_i = \beta D_{\rm m}\left[f_{\rm chem} - \kappa(x_i - x_{\rm c}) - f_{\rm max}\sin(2\pi x_i/\ell)\right] + \xi_i, \tag{4.20}$$

where $f_{\rm max} = \pi E^{\ddagger}/\ell$ is the maximum conservative force arising from the periodic potential. (Note that the motor energy landscapes here are parameterized differently from Ref. [41] to ensure consistency with Ref. [42]; the results of Fig. 4.5 are unchanged.) The cargo motion still obeys Eq. (4.1a). This continuous model, inspired by similar models of other molecular machines [14, 43], produces motor dynamics qualitatively similar to commonly used discrete-step models [17, 19]. In the limit as $f_{\rm max}/f_{\rm chem} \to 0$, this system is identical to the analytically tractable system (described by Eq. (4.1)) centrally featured in this chapter.

While these new equations of motion cannot be solved analytically, we can numerically simulate the dynamics of this system, by integrating the $N + 1$ Langevin equations for a given value of N. Obtaining full time-dependent probability distributions through simulation is computationally intractable for large N, so we compute only properties that depend solely on the average system dynamics. Figure 4.5 shows for this more complex model the scaling with N of the mean velocity $\langle v \rangle$, chemical power consumption $\dot{W}_{\rm chem}$, and Stokes efficiency $\eta_{\rm S}$, for a variety of barrier heights. (Calculating the coefficient of variation for large N is computationally intractable.) The scaling laws in the limiting regimes of large and small N (outlined in Table 4.1) still accurately reflect the limiting scaling behaviour for this generalization. As a direct result of the scaling laws generalizing, the performance trade-offs [Fig. 4.4a

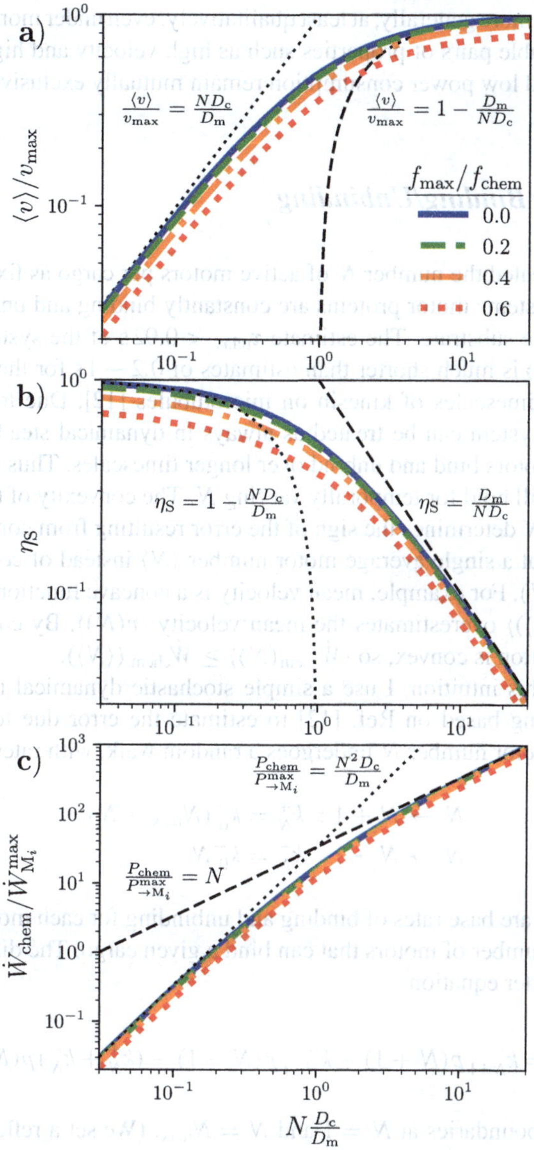

Fig. 4.5 Scaling of **a** the scaled mean velocity $\langle v \rangle / v_{\max}$, **b** the Stokes efficiency η_S, and **c** the scaled chemical power consumption $\dot{W}_{\text{chem}}/P^{\max}_{\to M_i}$, as a function of the motor number N scaled by the diffusivity ratio D_m/D_c. We simulate $N \in [1, 2, 3, \ldots, 1000]$ for several different ratios $f_{\max}/f_{\text{chem}}$ of the maximum conservative force to the chemical driving force. $f_{\max}/f_{\text{chem}} = 0$ (solid blue) recovers the analytically tractable model. Black curves indicate the scaling laws from Table 4.1 in the small-N (dotted) and large-N (dashed) limits. Parameters: $f_{\text{chem}} = 10$, $\kappa = 1$, $D_c = 0.03 D_m$; in dimensionless units chosen so that $D_m = \kappa = \ell = 1$

and c] also apply more generally, at least qualitatively: even under more general motor dynamics, desirable pairs of properties such as high velocity and high efficiency or high velocity and low power consumption remain mutually exclusive.

4.4.2 Motor Binding/Unbinding

So far, I have treated the number N of active motors per cargo as fixed, although in real transport systems motor proteins are constantly binding and unbinding to both the cargo and the substrate. The estimate $\tau_{\text{relax}} < 0.02\,\text{s}$ of the system's relaxation time (Sect. 4.3.1) is much shorter than estimates of $0.2 - 1\,\text{s}$ for the motor binding and unbinding timescales of kinesin on microtubules [12]. Due to this timescale separation, the system can be treated as always in dynamical steady state at fixed N even when motors bind and unbind over longer timescales. Thus the steady-state results should still hold for temporally varying N. The convexity of these properties with respect to N determines the sign of the error resulting from computing steady-state quantities at a single average motor number $\langle N\rangle$ instead of considering a full distribution $p(N)$. For example, mean velocity is a concave function of N, so when N varies $\langle v\left(\langle N\rangle\right)\rangle$ overestimates the mean velocity $\langle v(N)\rangle$. By contrast, the total power consumption is convex, so $\langle \dot{W}_{\text{chem}}(N)\rangle \geq \dot{W}_{\text{chem}}\left(\langle N\rangle\right)$.

To confirm this intuition, I use a simple stochastic dynamical model of motor binding/unbinding based on Ref. [12] to estimate the error due to treating N as constant. The motor number N undergoes a random walk, with rates

$$N \to N+1: \; k_N^+ = k_0^+(N_{\max} - N), \tag{4.21a}$$

$$N \to N-1: \; k_N^- = k_0^- N. \tag{4.21b}$$

Here k_0^+ and k_0^- are base rates of binding and unbinding for each motor, and $N_{\max}$ is the maximum number of motors that can bind a given cargo. The distribution $p(N)$ satisfies the master equation

$$\frac{\partial}{\partial t} p(N) = k_{N+1}^- p(N+1) + k_{N-1}^+ p(N-1) - (k_N^+ + k_N^-)p(N)\,, \tag{4.22}$$

with reflecting boundaries at $N = 1$ and $N = N_{\max}$. (We set a reflecting boundary at $N = 1$ rather than $N = 0$ because we are interested only in the behaviour of the system when there are motors attached.)

The master equation (4.22) with these boundary conditions has a time-independent steady-state solution. To simplify the analysis I take the limit $N_{\max} \to \infty$ and $k_0^+ \to 0$ such that $N_{\max} k_0^+ / k_0^- = \lambda$ is fixed. The steady-state solution is then

$$p(N) = \frac{\lambda^N}{N!(e^\lambda - 1)}, \tag{4.23}$$

defined for $N \geq 1$. This is a zero-truncated Poisson distribution [44], with mean

$$\langle N \rangle = \frac{\lambda}{1 - e^{-\lambda}}. \tag{4.24}$$

This distribution can then be used to calculate mean values of different steady-state properties of the transport system, averaged over $p(N)$. In particular, I estimate the error involved in assuming the system is well described by a constant (rather than fluctuating) number of motors, taking the mean velocity as an example.

For a given average number $\langle N \rangle$ of motors, we compare the mean velocity (4.6) evaluated at fixed $N = \langle N \rangle$ to the velocity instead averaged over the distribution $p(N)$ with the parameter λ chosen so that $\sum_{N=1}^{\infty} N p(N) = \langle N \rangle$. The error in mean velocity incurred by assuming fixed N is

$$\epsilon \equiv \frac{\left\langle v\left(\langle N \rangle_{p(N)}\right)\right\rangle_{\mathrm{ss}} - \left\langle \langle v(N) \rangle_{\mathrm{ss}} \right\rangle_{p(N)}}{\left\langle \langle v(N) \rangle_{\mathrm{ss}} \right\rangle_{p(N)}}, \tag{4.25}$$

where $\langle \cdot \rangle_{\mathrm{ss}}$ denotes an ensemble average over the system dynamics at fixed N, and $\langle \cdot \rangle_{p(N)}$ denotes an average over the probability distribution $p(N)$.

Figure 4.6 shows this error over a range of different values of $\langle N \rangle$ and $D_{\mathrm{c}}/D_{\mathrm{m}}$. The error resulting from assuming a fixed number of motors is less than 7%, and with other quantities suffering similar magnitudes of error. Note that the error is maximized for small $\langle N \rangle$, and for $\langle N \rangle D_{\mathrm{c}}/D_{\mathrm{m}} \approx 1$.

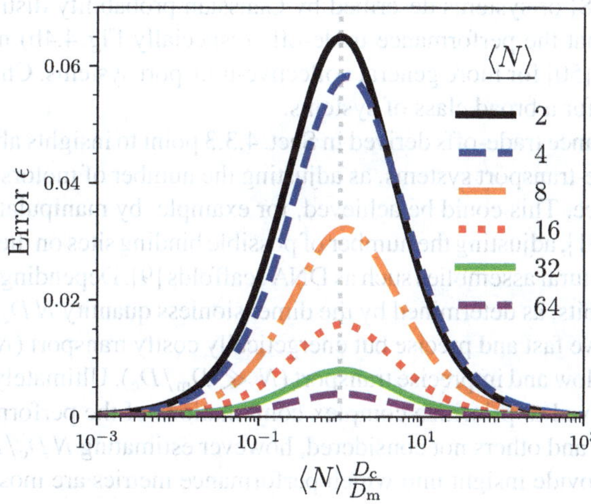

Fig. 4.6 Error (4.25) in mean velocity between fluctuating (described by Eq. (4.23)) and constant motor number, both with equal mean motor number $\langle N \rangle$, as a function of $\langle N \rangle D_{\mathrm{c}}/D_{\mathrm{m}}$. Vertical gray dotted line indicates $\langle N \rangle D_{\mathrm{c}}/D_{\mathrm{m}} = 1$

4.5 Discussion

In this chapter, I introduced a simple model for collective intracellular transport by an arbitrary number of transport motors. This model is stochastic, thermodynamically consistent, and can be solved analytically for arbitrary motor number N. Using this model, I derived analytic expressions for several steady-state properties, including dynamic properties such as mean velocity, effective diffusivity, and precision, as well as thermodynamic quantities such as power, heat, and efficiency. I found qualitatively different N-dependence for these quantities in the two opposing regimes of high and low N (compared to the motor-cargo diffusivity ratio $D_{\mathrm{m}}/D_{\mathrm{c}}$), summarized by simple scaling laws (Table 4.1). The model should best reflect the physics of motors whose energy-landscape features are smaller in scale than the magnitude of their chemical driving, however the numerical explorations in Sect. 4.4.1 suggest that many of this chapter's results generalize well beyond this regime.

I also examined trade-offs between several performance metrics that are expected to be generally important for real transport systems: many pairs of desirable properties, for example high velocity and high efficiency, are mutually exclusive in these systems. The trade-off between efficiency and velocity (Fig. 4.4c) has also been explored theoretically in other types of molecular machines [45, 46]. The incompatibility of high velocity and high efficiency seems to be a general feature of these types of systems, and has also been seen experimentally for myosin motors [30]. These findings are reminiscent of the thermodynamic uncertainty relation [47], which lower bounds the product of uncertainty and entropy production. At steady state the system saturates this bound, which may be a universal feature of linear systems with only one driving force [48] or systems described by Gaussian probability distributions [49]. This suggests that the performance trade-offs (especially Fig. 4.4b) may in fact be Pareto frontiers [50] for more general collective-transport systems. Chapter 5 proves this hypothesis for a broad class of systems.

The performance trade-offs derived in Sect. 4.3.3 point to insights about optimization in collective-transport systems, as adjusting the number of motors per cargo can tune performance. This could be achieved, for example, by manipulating the motor concentration [51], adjusting the number of possible binding sites on the cargo [52], or using extra structural assemblies such as DNA scaffolds [9]. Depending on the regime the system inhabits, as determined by the dimensionless quantity $N D_{\mathrm{c}}/D_{\mathrm{m}}$, systems can either achieve fast and precise but energetically costly transport ($N \gg D_{\mathrm{m}}/D_{\mathrm{c}}$), or efficient but slow and imprecise transport ($N \ll D_{\mathrm{m}}/D_{\mathrm{c}}$). Ultimately, real systems have likely evolved to prioritize complex combinations of the performance metrics considered here and others not considered, however estimating $N D_{\mathrm{c}}/D_{\mathrm{m}}$ for *in vivo* systems may provide insight into which performance metrics are most important in specific systems.

> **Design Principle:** The number N of motors in a collective transport system can be tuned to optimize different performance metrics of interest. $N \gg D_m/D_c$ maximizes velocity and precision at the cost of high power consumption and decreased efficiency, while $N \ll D_m/D_c$ maximizes efficiency but leads to slow and imprecise transport.

Few experiments have measured exact numbers of motors in collective-transport systems; nonetheless, motor numbers have indeed been measured, for example by identifying discrete peaks in a distinctly multimodal velocity distribution. Using this technique to study transport of beads by kinesin motors, Refs. [4, 5] found a mean velocity roughly proportional to the number of motors, consistent with my predictions for the small-N regime; the bead diameter and solution viscosity indicate that these experiments had $N < D_c/D_m$. Similar experimental investigations have also found velocity to be a concave function of N, qualitatively consistent with my predictions [53]. Other experiments have varied the concentration of motor proteins in solution, a rough proxy for the number of motors per cargo. Kinesin motors attached to a substrate while transporting long microtubules [54, 55] produce velocity that is a concave function of motor concentration, as this chapter predicts; similar results have been found for myosin motors pulling actin filaments [7].

Using models similar to ours, but with discrete steps rather than continuous motor dynamics, Refs. [16, 19] found a similar monotonic and concave functional dependence of the mean velocity $\langle v \rangle$ on motor number N, although their findings are restricted to relatively small N. Likewise, Ref. [19] also found $D_{\text{eff}} \propto 1/N$ for the greatest N they investigated. Another recent study [20] found that the total load capacity, or effective stall force, is N times the stall force for a single motor. The model can easily incorporate a constant external force f_{ext} pulling the cargo in the opposite direction of the chemical force driving the motors. This simply requires adding a term $-\beta D_c f_{\text{ext}}$ to the left side of (4.1a), preserving analytic solubility of the dynamical equations. The mean velocity in this case is

$$\langle v \rangle_{f_{\text{ext}}} = \langle v \rangle_0 \left(1 - \frac{f_{\text{ext}}}{N f_{\text{chem}}} \right), \tag{4.26}$$

where $\langle v \rangle_0$ is the mean velocity for $f_{\text{ext}} = 0$ given by Eq. (4.6). From this we identify the stall force $f_s = N f_{\text{chem}}$ which scales linearly with the number of motors, independent of the diffusivity ratio.

The model also ignores other possibly relevant effects such as motor-motor interactions and discretization of motor steps. Depending on the time- and lengthscales of resolution, motor proteins like kinesin can be thought of as taking discrete steps [56], in contrast to the continuous motion assumed here. Nevertheless, treating the motor dynamics as continuous should be valid as long as the relaxation timescale for the motors is sufficiently separated from that of the cargo (either much larger or much smaller) so that the steady-state separation distance distribution (4.5) converges to the resulting steady-state distribution. This assumption is valid for both $N \gg D_m/D_c$

and $N \ll D_m/D_c$, but may break down for $N \approx D_m/D_c$, where system behaviour may be more sensitive to the exact motor dynamics. Interactions between motors may become relevant under certain conditions. Computational studies indicate the possibility of long-range cooperative interactions between kinesin motors through microtubules [57] as well as crowding effects such as traffic jams when large numbers of motors are present [58]. While this chapter has not incorporated these effects, generalizing the results using model-agnostic considerations from the theory of stochastic thermodynamics is a promising future direction. In the next chapter I will consider a more general class of collective transport models.

References

1. Hancock WO (2008) Intracellular transport: kinesins working together. Curr Biol 18(16):R715–R717
2. Leopold PL, McDowall AW, Pfister KK, Bloom GS, Brady ST (1992) Association of kinesin with characterized membrane-bounded organelles. Cell Motil Cytoskelet 23(1):19–33
3. Hatsumi M, Endow SA (1992) Mutants of the microtubule motor protein, nonclaret disjunctional, affect spindle structure and chromosome movement in meiosis and mitosis. J Cell Sci 101(3):547–559
4. Shtridelman Y, Cahyuti T, Townsend B, DeWitt D, Macosko JC (2008) Force–velocity curves of motor proteins cooperating in vivo. Cell Biochem Biophys 52(1):19–29
5. Shtridelman Y, Holzwarth GM, Bauer CT, Gassman NR, DeWitt DA, Macosko JC (2009) In vivo multimotor force–velocity curves by tracking and sizing sub-diffraction limited vesicles. Cell Mol Bioeng 2(2):190–199
6. Svoboda K, Block SM (1994) Force and velocity measured for single kinesin molecules. Cell 77(5):773–784
7. Rastogi K, Puliyakodan MS, Pandey V, Nath S, Elangovan R (2016) Maximum limit to the number of myosin II motors participating in processive sliding of actin. Sci Rep 6(1):1–10
8. DelRosso NV, Derr ND (2017) Exploiting molecular motors as nanomachines: the mechanisms of de novo and re-engineered cytoskeletal motors. Curr Opin Biotechnol 46:20–26
9. Furuta K, Furuta A, Toyoshima YY, Amino M, Oiwa K, Kojima H (2013) Measuring collective transport by defined numbers of processive and nonprocessive kinesin motors. Proc Natl Acad Sci 110(2):501–506
10. Derr ND, Goodman BS, Jungmann R, Leschziner AE, Shih WM, Reck-Peterson SL (2012) Tug-of-war in motor protein ensembles revealed with a programmable DNA origami scaffold. Science 338(6107):662–665
11. Hunt AJ, Gittes F, Howard J (1994) The force exerted by a single kinesin molecule against a viscous load. Biophys J 67(2):766–781
12. Klumpp S, Lipowsky R (2005) Cooperative cargo transport by several molecular motors. Proc Natl Acad Sci 102(48):17284–17289
13. Kunwar A, Mogilner A (2010) Robust transport by multiple motors with nonlinear force-velocity relations and stochastic load sharing. Phys Biol 7(1):016012
14. Zimmermann E, Seifert U (2015) Effective rates from thermodynamically consistent coarse-graining of models for molecular motors with probe particles. Phys Rev E 91(2):022709
15. Brown AI, Sivak DA (2019a) Pulling cargo increases the precision of molecular motor progress. EPL (Eur Lett) 126(4):40004
16. Korn CB, Klumpp S, Lipowsky R, Schwarz US (2009) Stochastic simulations of cargo transport by processive molecular motors. J Chem Phys 131(24):12B624
17. McKinley SA, Athreya A, Fricks J, Kramer PR (2012) Asymptotic analysis of microtubule-based transport by multiple identical molecular motors. J Theor Biol 305:54–69

18. Li X, Lipowsky R, Kierfeld J (2012) Critical motor number for fractional steps of cytoskeletal filaments in gliding assays. PLoS One 7(8):e43219
19. Bhat D, Gopalakrishnan M (2016) Transport of organelles by elastically coupled motor proteins. Eur Phys J E 39(7):1–13
20. Bhat D, Gopalakrishnan M (2017) Stall force of a cargo driven by N interacting motor proteins. EPL (Eur Lett) 117(2):28004
21. Arpağ G, Shastry S, Hancock WO, Tüzel E (2014) Transport by populations of fast and slow kinesins uncovers novel family-dependent motor characteristics important for in vivo function. Biophys J 107(8):1896–1904
22. Brown AI, Sivak DA (2019b) Theory of nonequilibrium free energy transduction by molecular machines. Chem Rev 120(1):434–459
23. Kawaguchi K, Uemura S, Ishiwata S (2003) Equilibrium and transition between single- and double-headed binding of kinesin as revealed by single-molecule mechanics. Biophys J 84(2):1103–1113
24. Veigel C, Schmitz S, Wang F, Sellers JR (2005) Load-dependent kinetics of myosin-V can explain its high processivity. Nat Cell Biol 7(9):861–869
25. Visscher K, Schnitzer MJ, Block SM (1999) Single kinesin molecules studied with a molecular force clamp. Nature 400(6740):184–189
26. Debold EP, Patlak JB, Warshaw DM (2005) Slip sliding away: load-dependence of velocity generated by skeletal muscle myosin molecules in the laser trap. Biophys J 89(5):L34–L36
27. Schnitzer MJ, Block SM (1997) Kinesin hydrolyses one ATP per 8-nm step. Nature 388(6640):386–390
28. Milo R, Phillips R (2015) Cell Biology by the Numbers. Garland Science
29. Fallesen TL, Macosko JC, Holzwarth G (2011) Force–velocity relationship for multiple kinesin motors pulling a magnetic bead. Eur Biophys J 40(9):1071–1079
30. Purcell TJ, Naber N, Franks-Skiba K, Dunn AR, Eldred CC, Berger CL, Málnási-Csizmadia A, Spudich JA, Swank DM, Pate E et al (2011) Nucleotide pocket thermodynamics measured by EPR reveal how energy partitioning relates myosin speed to efficiency. J Mol Biol 407(1):79–91
31. Vu HT, Chakrabarti S, Hinczewski M, Thirumalai D (2016) Discrete step sizes of molecular motors lead to bimodal non-Gaussian velocity distributions under force. Phys Rev Lett 117(7):078101
32. Ahmed WW, Saif TA (2014) Active transport of vesicles in neurons is modulated by mechanical tension. Sci Rep 4(1):1–7
33. Luby-Phelps K (1999) Cytoarchitecture and physical properties of cytoplasm: volume, viscosity, diffusion, intracellular surface area. Int Rev Cytol 192:189–221
34. Risken H (1996) Fokker-Planck equation. Springer
35. Toprak E, Yildiz A, Hoffman MT, Rosenfeld SS, Selvin PR (2009) Why kinesin is so processive. Proc Natl Acad Sci 106(31):12717–12722
36. Nakul U, Gopalakrishnan M (2021) Frictional drag produced by motor proteins during cargo transport. EPL (Eur Lett) 133(6):68002
37. Leibler S, Huse DA (1993) Porters versus rowers: a unified stochastic model of motor proteins. J Cell Biol 121(6):1357–1368
38. Brown CE (1998) Coefficient of variation. In: Applied multivariate statistics in geohydrology and related sciences. Springer, pp 155–157
39. de Groot SR, Mazur P (1969) Non-equilibrium thermodynamics. North-Holland Publication Co, Amsterdam
40. Pietzonka P, Barato AC, Seifert U (2016) Universal bound on the efficiency of molecular motors. J Stat Mech Theory Exp 2016(12):124004
41. Leighton MP, Sivak DA (2022a) Performance scaling and trade-offs for collective motor-driven transport. New J Phys 24(1):013009
42. Leighton MP, Sivak DA (2022b) Dynamic and thermodynamic bounds for collective motor-driven transport. Phys Rev Lett 129(11):118102
43. Lathouwers E, Lucero JNE, Sivak DA (2020) Nonequilibrium energy transduction in stochastic strongly coupled rotary motors. J Phys Chem Lett 11(13):5273–5278

44. Johnson NL, Kemp AW, Kotz S (2005) Univariate discrete distributions, vol 444. John Wiley & Sons
45. Wagoner JA, Dill KA (2019) Opposing pressures of speed and efficiency guide the evolution of molecular machines. Mol Biol Evol 36(12):2813–2822
46. Wagoner JA, Dill KA (2021) Evolution of mechanical cooperativity among myosin II motors. Proc Natl Acad Sci 118(20)
47. Horowitz JM, Gingrich TR (2020) Thermodynamic uncertainty relations constrain non-equilibrium fluctuations. Nat Phys 16(1):15–20
48. Barato AC, Seifert U (2015) Thermodynamic uncertainty relation for biomolecular processes. Phys Rev Lett 114(15):158101
49. Saryal S, Friedman HM, Segal D, Agarwalla BK (2019) Thermodynamic uncertainty relation in thermal transport. Phys Rev E 100(4):042101
50. Shoval O, Sheftel H, Shinar G, Hart Y, Ramote O, Mayo A, Dekel E, Kavanagh K, Alon U (2012) Evolutionary trade-offs, Pareto optimality, and the geometry of phenotype space. Science 336(6085):1157–1160
51. Ndlec FJ, Surrey T, Maggs AC, Leibler S (1997) Self-organization of microtubules and motors. Nature 389(6648):305–308
52. Kumar J, Yu H, Sheetz MP (1995) Kinectin, an essential anchor for kinesin-driven vesicle motility. Science 267(5205):1834–1837
53. Gagliano J, Walb M, Blaker B, Macosko JC, Holzwarth G (2010) Kinesin velocity increases with the number of motors pulling against viscoelastic drag. Eur Biophys J 39(5):801–813
54. Kaneko T, Ando S, Furuta K, Oiwa K, Shintaku H, Kotera H, Yokokawa R (2019) Transport of microtubules according to the number and spacing of kinesin motors on gold nano-pillars. Nanoscale 11(20):9879–9887
55. Howard J, Hudspeth AJ, Vale RD (1989) Movement of microtubules by single kinesin molecules. Nature 342(6246):154–158
56. Svoboda K, Schmidt CF, Schnapp BJ, Block SM (1993) Direct observation of kinesin stepping by optical trapping interferometry. Nature 365(6448):721–727
57. Wijeratne SS, Fiorenza SA, Neary AE, Subramanian R, Betterton MD (2022) Motor guidance by long-range communication on the microtubule highway. Proc Natl Acad Sci 119(28):e2120193119
58. Klumpp S, Nieuwenhuizen TM, Lipowsky R (2005) Movements of molecular motors: Ratchets, random walks and traffic phenomena. Phys E Low Imensional Syst Nanostructures 29(1–2):380–389

Chapter 5
Dynamic and Thermodynamic Bounds for Collective Motor-Driven Transport

Molecular motors work collectively to transport cargo within cells, with anywhere from one to several hundred motors towing a single cargo. For a broad class of collective-transport systems, I use the Jensen lower bound for the entropy production rate to derive constraints on performance. This implies new bounds on the velocity, efficiency, and precision of general transport systems and a set of analytic Pareto frontiers for identical motors. In a specific model, I identify conditions for saturation of these Pareto frontiers.

5.1 Introduction

Many specific models of transport systems have been explored, including deterministic phenomenological models [1–3], discrete stochastic models [4–6], and continuous stochastic models like the one in the last chapter [7, 8]. A common goal of these investigations has been to determine how various parameters (such as coupling strength, stall force, diffusivity, and number of motors) tune the performance of these systems. Performance metrics of interest include dynamical quantities such as velocity and precision, and thermodynamic quantities like efficiency and power consumption [9]. While the behaviour of specific model systems has been explored, considerably less is known about the fundamental performance limits for transport systems in general, agnostic of model details.

The behaviour of transport systems is restricted by two fundamental thermodynamic limitations. First and foremost, they must obey the second law of thermodynamics, the most useful form in these contexts stating that at steady state the ensemble-averaged rate of global entropy production cannot be negative [10]. Second, the recently established Thermodynamic Uncertainty Relation (TUR) [11–13] lower bounds products of the entropy production rate and uncertainties in various

M. Leighton, *Stochastic Thermodynamics of Multicomponent Molecular Machines*, Springer Theses, https://doi.org/10.1007/978-3-032-02204-2_5

currents at steady state. These key inequalities have been used to derive bounds on various performance metrics, for example efficiency [14].

In this chapter, I consider the thermodynamics of motor-driven intracellular transport in far more generality than in Chap. 4. I apply the Jensen bound, derived in Chap. 3, and the TUR to derive a set of bounds on performance metrics such as velocity, efficiency, and precision. These bounds constrain emergent properties of collective systems for arbitrary numbers of motors of any directionality, using only bare properties of individual subsystems. This theory holds for a broad class of collective-transport systems, independent of any model-specific interaction potentials or spatially inhomogeneous energy landscapes. For identical motors, I then derive analytic expressions for several Pareto frontiers constraining combinations of performance metrics. Finally, I simulate an example system to illustrate these bounds and conditions sufficient for their saturation.

5.2 Theory and Model

As in Chap. 4, consider N transport motors coupled to a diffusing molecular cargo, all moving in one dimension. Each motor interacts with the cargo via a molecular linker, and is characterized by a mechanochemical cycle through which it transduces chemical power into directed forward motion. The cargo undergoes overdamped Brownian motion (with bare diffusivity D_c) constrained by interactions with each motor (with its own bare diffusivity D_i).

Each motor in isolation experiences a constant chemical driving force f_i, along with a spatially periodic potential-energy landscape arising due to interactions with the substrate it walks along (e.g., microtubules for kinesin). This may include features such as metastable states and energy barriers. (Multiple cargos are trivially incorporated as motors with no chemical driving force, $f_i = 0$.) I assume that each motor tightly couples its chemical and mechanical degrees of freedom, consistent with experiments on kinesin and myosin motors [15–17]. Motors and cargo are coupled via the total potential energy $V(\boldsymbol{x})$ for the vector $\boldsymbol{x} \equiv \{x_\mathrm{c}, x_1, ..., x_i, ..., x_N\}$ of cargo position x_c and motor positions $\{x_i\}_{i=1}^N$. This potential describes, e.g., the molecular linkers attaching each motor to the cargo and attractive or repulsive interactions between motors. The dynamics of this model are given by the multipartite overdamped Langevin and Fokker-Planck equations, Eqs. (2.56) and (2.59) from Sect. 2.5.

In the steady-state limit the subsystems (cargo and motors) must stay together, i.e., the relative coordinates $x_i - x_\mathrm{c}$ reach time-independent distributions at steady state so that each subsystem has the same mean velocity, $\langle v \rangle \equiv \langle \dot{x}_\mathrm{c} \rangle = \langle \dot{x}_i \rangle$. In terms of the potential $V(\boldsymbol{x})$, this requires that all subsystems are coupled and at long distances any repulsive interactions are dominated by attraction.

5.2.1 Performance Metrics

I consider several performance metrics for collective-transport systems, all of which were introduced in Chap. 4. These include the mean velocity:

$$\langle v\rangle = \left\langle \frac{J_{\rm c}(x,t)}{P(x,t)} \right\rangle \tag{5.1a}$$

$$= \left\langle \frac{J_i(x,t)}{P(x,t)} \right\rangle, \tag{5.1b}$$

the average rate of total chemical-energy consumption by the N motors:

$$\dot{W}_{\rm chem} \equiv \sum_{i=1}^{N} \langle f_i \dot{x}_i \rangle \tag{5.2a}$$

$$= \sum_{i=1}^{N} f_i \langle v\rangle, \tag{5.2b}$$

the Stokes efficiency:

$$\eta_{\rm S} \equiv \frac{\zeta_{\rm c}\langle v\rangle^2}{\dot{W}_{\rm chem}}, \tag{5.3}$$

the effective diffusivity (4.7), and the precision (quantified by the coefficient of variation):

$$\theta \equiv \frac{\sqrt{\langle \delta x_{\rm c}^2\rangle}}{\langle x_{\rm c}\rangle}. \tag{5.4}$$

Here $\langle \delta x_{\rm c}^2\rangle$ is the variance of the cargo position $x_{\rm c}$.

From Sect. 2.5, the average rates of dimensionless entropy production for each subsystem are [18]

$$\dot{\Sigma}_i = \frac{1}{D_i}\left\langle \left[\frac{J_i(x,t)}{P(x,t)}\right]^2 \right\rangle \geq 0 \tag{5.5a}$$

$$\dot{\Sigma}_{\rm c} = \frac{1}{D_{\rm c}}\left\langle \left[\frac{J_{\rm c}(x,t)}{P(x,t)}\right]^2 \right\rangle \geq 0. \tag{5.5b}$$

The total entropy production rate is their sum, $\dot{\Sigma} = \dot{\Sigma}_{\rm c} + \sum_{i=1}^{N} \dot{\Sigma}_i$. For a diffusive cargo with no external forces, the entropy production equals the total chemical power:

$$\dot{\Sigma} = \beta \dot{W}_{\rm chem}. \tag{5.6}$$

5.3 Bounds for General Systems

Applying the Jensen bounds to this system yields three inequalities constraining the partial and total entropy production rates:

$$\dot{\Sigma}_i \geq \frac{1}{D_i}\langle v\rangle^2, \tag{5.7a}$$

$$\dot{\Sigma}_{\mathrm{c}} \geq \frac{1}{D_{\mathrm{c}}}\langle v\rangle^2, \tag{5.7b}$$

$$\dot{\Sigma} \geq \frac{1}{D_{\mathrm{bare}}}\langle v\rangle^2. \tag{5.7c}$$

Here D_{bare} is the "bare collective diffusivity", the inverse of the total friction coefficient from summing the individual friction coefficients (inversely proportional to bare diffusivities) of each subsystem:

$$D_{\mathrm{bare}} \equiv \left(\frac{1}{D_{\mathrm{c}}} + \sum_{i=1}^{N} \frac{1}{D_i}\right)^{-1}. \tag{5.8}$$

Physically, D_{bare} is the effective diffusivity under a potential that depends only on relative subsystem positions.

Equation (5.7c) constitutes a general, model-independent, lower bound (non-negative and thus tighter than the second law (5.5)) on the entropy production required for a collective-transport system with N motors to maintain mean velocity $\langle v\rangle$.

The collective-transport system is also constrained by the long-time limit of the thermodynamic uncertainty relation (TUR) [11–13], whose most useful form for this system is

$$\dot{\Sigma} t \, \frac{\langle \delta x_{\mathrm{c}}^2\rangle}{\langle x_{\mathrm{c}}\rangle^2} \geq 2. \tag{5.9}$$

Identifying $\langle v\rangle = \langle x_{\mathrm{c}}\rangle / t$ and D_{eff} (4.7) recasts this inequality as

$$\dot{\Sigma} \geq \frac{1}{D_{\mathrm{eff}}}\langle v\rangle^2, \tag{5.10}$$

which has the same form as the Jensen bound (5.7c). Equations (5.7c) and (5.10) thus constitute two bounds on the entropy production. In general, either of these bounds can be tighter. Even for a single particle in a tilted sinusoidal potential, either $D_{\mathrm{bare}} < D_{\mathrm{eff}}$ or $D_{\mathrm{bare}} > D_{\mathrm{eff}}$ is possible, depending on the ratio of the barrier height to the driving force [19].

Substituting (5.6) and the Stokes efficiency (5.3) into (5.7c) gives an upper bound on η_{S}:

$$\eta_{\mathrm{S}} \leq \frac{D_{\mathrm{bare}}}{D_{\mathrm{c}}}. \tag{5.11}$$

This is similar, but not equivalent, to a previous bound [14]: $\eta_S \leq D_{\text{eff}}/D_c$. Like the Jensen bound (5.7c) and TUR (5.10), either of these two bounds can be tighter in different circumstances.

Likewise, substituting (5.6) and $\dot{W}_{\text{chem}} = f_{\text{tot}}\langle v\rangle$ (for total force $f_{\text{tot}} = \sum_{i=1}^{N} f_i$ which I assume without loss of generality to be non-negative) into (5.7c) yields an upper bound on the average velocity:

$$\langle v\rangle \leq \beta D_{\text{bare}} f_{\text{tot}}. \tag{5.12}$$

Finally, substituting (5.6) and the coefficient of variation (5.4) into the TUR (5.9) and employing the velocity inequality (5.12) gives an upper bound on the precision through a lower bound on the coefficient of variation:

$$\theta \geq \frac{1}{\beta f_{\text{tot}}}\sqrt{\frac{2}{D_{\text{bare}} t}} \tag{5.13}$$

These three bounds (5.11)–(5.13) constitute this chapter's second major result, constraining global system properties using only properties (D_{bare}, D_c, and f_{tot}) of each individual subsystem in isolation.

5.4 Identical Motors

I illustrate the utility of these performance bounds with the special case where transport motors are identical, each with diffusivity D_m and driving force f_{chem}. This reflects many biological systems of interest, such as identical kinesin motors towing a large vesicle, or identical myosin motors pulling an actin filament. The Jensen bound (5.7c) becomes

$$\Sigma \geq \left(\frac{1}{D_c} + \frac{N}{D_m}\right)\langle v\rangle^2. \tag{5.14}$$

The general performance bounds (5.11)–(5.13) can be rewritten in terms of more natural variables as

$$\eta_S \leq \left(1 + \frac{N D_c}{D_m}\right)^{-1}, \tag{5.15a}$$

$$\langle v\rangle \leq \beta N D_c f_{\text{chem}} \left(1 + \frac{N D_c}{D_m}\right)^{-1}, \tag{5.15b}$$

$$\theta \geq \frac{1}{\beta N f_{\text{chem}}}\sqrt{\frac{2}{D_c t}}\left(1 + \frac{N D_c}{D_m}\right)^{1/2}. \tag{5.15c}$$

Since $N D_c/D_m > 0$, a looser upper bound on the mean velocity is

$$v_{\max} = \beta D_m f_{\text{chem}}, \tag{5.16}$$

the mean velocity of a single motor in a flat potential subject to constant force f_{chem}. Likewise, since $N \geq 1$, the Stokes efficiency has a looser upper bound of $(1 + D_c/D_m)^{-1}$.

Combining (5.15a) with (5.15b) gives a Pareto frontier between the Stokes efficiency and scaled mean velocity:

$$\eta_S + \frac{\langle v \rangle}{v_{\max}} \leq 1. \tag{5.17}$$

Similarly, combining (5.14) and (5.12) gives

$$\frac{\dot{W}_{\text{chem}}}{\dot{W}_i^{\max}} \geq \frac{D_m}{D_c} \frac{(\langle v \rangle/v_{\max})^2}{1 - \langle v \rangle/v_{\max}}, \tag{5.18}$$

a Pareto frontier constraining velocity and power consumption. Here $\dot{W}_i^{\max} = f_{\text{chem}} v_{\max}$ is the mean power consumption of a single motor at maximum velocity. These two Pareto frontiers follow solely from the Jensen bound (5.7c); the TUR (5.9) alone gives a Pareto frontier for power consumption and precision:

$$\beta \dot{W}_{\text{chem}} \theta^2 \geq 2. \tag{5.19}$$

5.4.1 External Force on Cargo

So far, the cargo has only encountered resistance from viscous drag; similar considerations also constrain performance for an additional external force f_{ext} on the cargo, in the direction opposite to f_{chem}. The entropy production rate is then

$$\begin{aligned} \dot{\Sigma} &= \beta(N f_{\text{chem}} - f_{\text{ext}})\langle v \rangle \\ &\geq \left(\frac{1}{D_c} + \frac{N}{D_m}\right) \langle v \rangle^2, \end{aligned} \tag{5.20}$$

Here, thermodynamic efficiency $\eta_T \equiv f_{\text{ext}}/(N f_{\text{chem}})$ is positive. Applying the Jensen bound leads to a Pareto frontier for thermodynamic efficiency and mean velocity:

$$\eta_T + \frac{D_m}{N D_{\text{bare}}} \frac{\langle v \rangle}{v_{\max}} \leq 1. \tag{5.21}$$

Since $D_m/ND_{bare} \geq 1$, a looser bound analogous to (5.17) is

$$\eta_T + \frac{\langle v \rangle}{v_{max}} \leq 1. \tag{5.22}$$

5.5 Example System

Consider an example with tunable parameters that can saturate the derived bounds. Each motor has periodic potential $V_i(x_i) = \frac{1}{2}E^\ddagger \cos(2\pi x_i/\ell)$ with barrier height $E^\ddagger$, period ℓ, and maximum conservative force $f_{max} = E^\ddagger/(2\ell)$. Each motor is linked to the cargo by a Hookean spring with spring constant κ and zero rest length [20], $U_i(x_c, x_i) = \frac{1}{2}\kappa(x_i - x_c)^2$. The motors do not directly interact. The total system potential is thus

$$V(\boldsymbol{x}) = \sum_{i=1}^{N} \left[\frac{1}{2}E^\ddagger \cos(2\pi x_i/\ell) + \frac{1}{2}\kappa(x_i - x_c)^2 \right]. \tag{5.23}$$

Equating the Kramers rate [22] for a single uncoupled motor hopping between adjacent landscape minima with experimentally measured rates for kinesin-1 motors [23] (Appendix A.2 gives details) yields $f_{max}/f_{chem} \approx 0.4$, which sets the scale of the parameter sweep.

Figure 5.1 illustrates that for $N = 2$ motors the dynamics change significantly as the barrier height increases. For $f_{max}/f_{chem} \ll 1$, the motors move continuously, while for $f_{max}/f_{chem} \gtrsim 1$ the motors hop between distinct states.

Figure 5.2 shows for $N = 2$ motors the trade-off between Stokes efficiency and velocity due to parametric variation of the diffusivity ratio D_c/D_m, for different barrier heights. When the motors face no barriers ($f_{max}/f_{chem} = 0$), the system exactly saturates the Pareto frontier (5.17). As f_{max}/f_{chem} increases, the performance trade-off degrades, falling increasingly far from the Pareto frontier.

While the η_S–$\langle v \rangle$ curve is linear for $f_{max}/f_{chem} = 0$, as f_{max}/f_{chem} increases it becomes increasingly convex. This suggests that for large energy barriers high efficiency or high velocity are more easily achieved than a compromise between the two. As expected, the velocity in the $ND_c/D_m \to \infty$ limit is exactly that of a single uncoupled motor on the same energy landscape, while the Stokes efficiency is zero. In the limit as $ND_c/D_m \to 0$, the velocity approaches zero and the Stokes efficiency approaches unity.

Figure 5.3 shows the trade-off between power consumption and velocity due to parametric variation of the motor number N and barrier heights. Since this Pareto frontier (5.18) depends on D_c/D_m, that ratio is held constant. Computational constraints limit simulations to small N. When the motors face no barriers ($f_{max}/f_{chem} = 0$), the system exactly saturates the Pareto frontier (5.18). As f_{max}/f_{chem} increases, the performance trade-off degrades, falling increasingly far from the Pareto frontier.

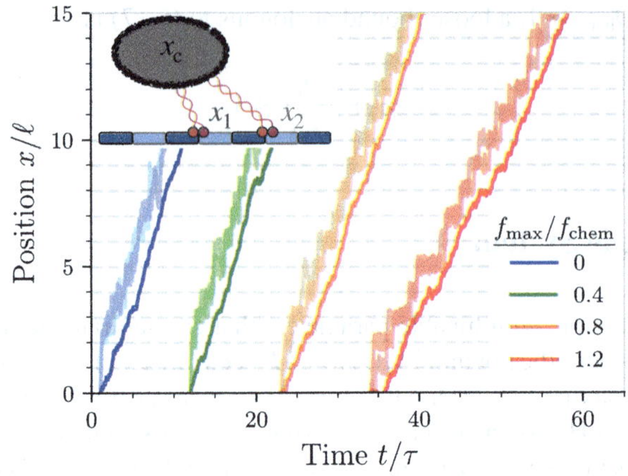

Fig. 5.1 Motor and cargo trajectories for the example system with $N = 2$ motors, for different f_{max}/f_{chem}. Dark curves: cargo; lighter curves: two motors. Dashed grey horizontal lines: local minima of motor potential energy. The start times of different trajectories are staggered for clarity. Position and time are respectively scaled by ℓ and $\tau = \ell^2/D_m$. Parameters used are $\beta f_{chem}\ell = 15$, $\beta\kappa\ell^2 = 7$, and $D_c/D_m = 1/30$. Copyrighted by the American Physical Society [21]

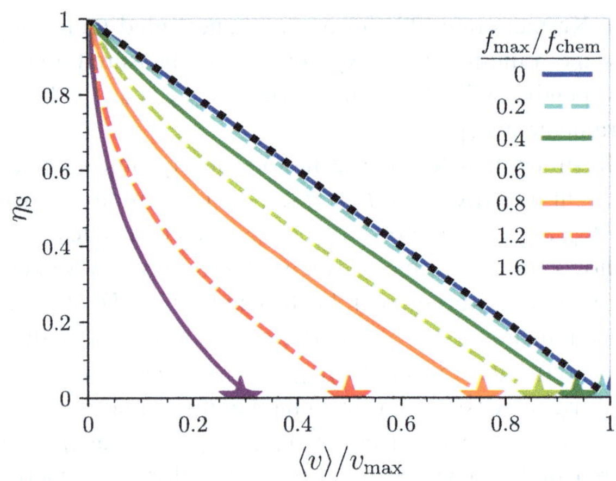

Fig. 5.2 Trade-off between Stokes efficiency η_S and scaled velocity $\langle v\rangle/v_{max}$ in the example system with $N = 2$ motors, plotted parametrically for $D_c/D_m = 10^{-3} - 10^3$. Colours: different f_{max}. Black dotted line: Pareto frontier (5.17). Stars: single uncoupled motor. Other parameters same as Fig. 5.1. Copyrighted by the American Physical Society [21]

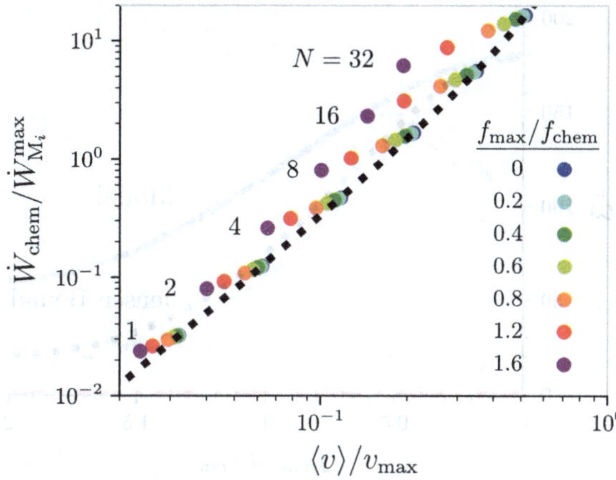

Fig. 5.3 Trade-off between power consumption $\dot{W}_{\text{chem}}$ and scaled velocity $\langle v\rangle/v_{\max}$ in the example system, plotted parametrically for $N = \{1, 2, 4, 8, 16, 32\}$ and different $f_{\max}$ (colours). Black dotted curve: Pareto frontier (5.18). Other parameters same as Fig. 5.1: $\beta f_{\text{chem}}\ell = 15$, $\beta\kappa\ell^2 = 7$, and $D_c/D_m = 1/30$. Uncertainties are smaller than the widths of the points. Copyrighted by the American Physical Society [21]

5.6 Comparison of Entropy Production Bounds

Figure 5.4 shows the entropy production rate of the specific model considered, along with the different lower bounds discussed: the Jensen bound (5.7c), TUR (5.10), and second law ($\dot{\Sigma} \geq 0$). The Jensen bound is generally the tightest constraint for my best estimates of reasonable model parameters in kinesin-vesicle systems.

More generally, the Jensen bound is tighter whenever $D_{\text{eff}} > D_{\text{bare}}$. I numerically explore the ratio $D_{\text{eff}}/D_{\text{bare}}$ over a 2D region of parameter space in Fig. 5.5, expanding around my best estimates of $\beta\kappa\ell^2 \approx 7$ and $f_{\max}/f_{\text{chem}} \approx 0.4 - 0.8$ for kinesin motors pulling vesicles, and find that $D_{\text{eff}} > D_{\text{bare}}$ (the Jensen bound is tighter) over a wide range of coupling strengths and barrier heights. For sufficiently large energy barriers and motor-cargo coupling, however, $D_{\text{eff}} < D_{\text{bare}}$ and thus the TUR is tighter. This is consistent with a previous study of coupled Brownian particles diffusing in a single periodic potential [24]. At high coupling strengths, subsystems can only cross energy barriers simultaneously [25], making forward progress only with much larger fluctuations whose rarity leads to decreased effective diffusivity. Likewise, high energy barriers could lead to phenomena like hindered diffusion, which lowers the effective diffusivity [26]. (Recall that any details of interactions with other subsystems or the substrate only affect D_{eff}, with D_{bare} uniquely determined by the diffusion coefficients of the components making up the system.)

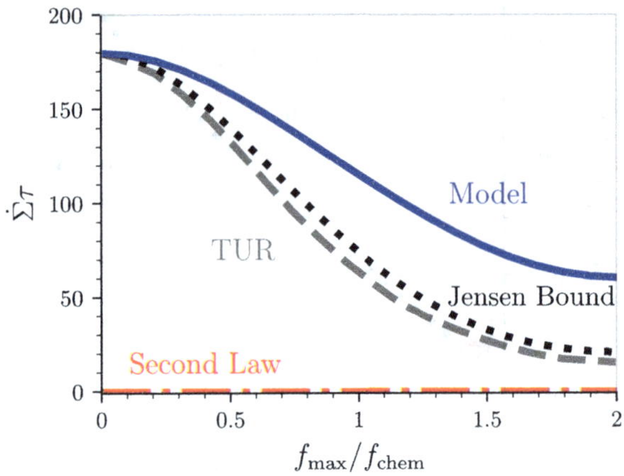

Fig. 5.4 Comparison of model entropy production with various lower bounds. Entropy production during time $\tau \equiv \ell^2/D_{\mathrm{m}}$ of the specific model considered in the main text (blue solid curve), the Jensen bound (5.7c) (black dotted), the TUR (5.10) (gray dashed), and the second law (red dot-dashed), each as a function of $f_{\max}/f_{\mathrm{chem}}$. Uncertainties are smaller than the widths of the curves. Parameters are $N = 2$ motors, $\beta f_{\mathrm{chem}}\ell = 15$, $\beta\kappa\ell^2 = 7$, and $D_{\mathrm{c}}/D_{\mathrm{m}} = 1/3$. Copyrighted by the American Physical Society [21]

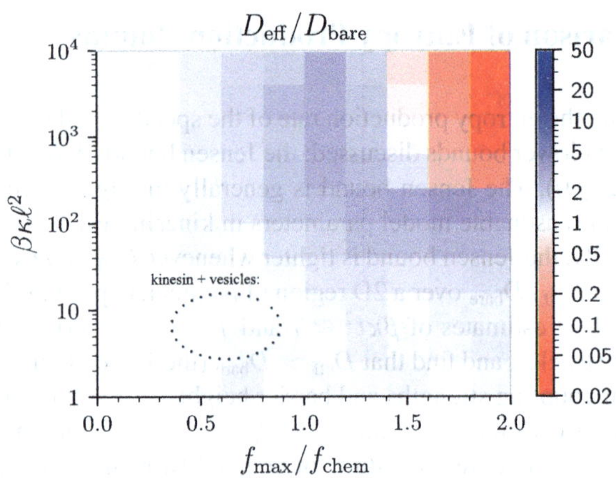

Fig. 5.5 The ratio $D_{\mathrm{eff}}/D_{\mathrm{bare}}$ between the effective diffusivity and the bare collective diffusivity, as a function of dimensionless coupling strength $\beta\kappa\ell^2$ and $f_{\max}/f_{\mathrm{chem}}$. Parameters are $N = 2$ motors, $\beta f_{\mathrm{chem}}\ell = 15$, and $D_{\mathrm{c}}/D_{\mathrm{m}} = 1/3$. Standard errors of the mean are each ~1–5%. Copyrighted by the American Physical Society [21]

5.7 Comparison with Experiments

Figure 5.6 shows experimental measurements of velocity and efficiency for myosin motors in several different animal tissues from Ref. [27]. For maximum velocity $v_{\max} = 12\,\mu\text{m/s}$ (to my knowledge, the highest observed in animal muscle tissue [28]), the predicted Pareto frontier (5.22) indeed bounds the experimentally observed performance. The assumption of a global $v_{\max}$ across many different species is reasonable so long as the difference between species-specific myosin motors comes predominantly from different potentials $V(x)$ as opposed to differences in the chemical driving force and bare diffusivity.

Figure 5.6 also shows experimental data for kinesin motors pulling diffusive cargo from Ref. [2], which measured average velocity for single transport motors (motor number inferred from multimodal velocity distributions) pulling vesicles, as a function of the vesicle diameter. Cargo diffusivity D_c is estimated from the measured diameter using the Stokes-Einstein relation [29] and reported measurements of temperature and viscosity, while D_m is fit to the theoretical prediction $\langle v\rangle/v_{\max} = (1 + D_m/D_c)^{-1}$ from Chap. 4, with $v_{\max} = 2\,\mu\text{m/s}$ (the maximum velocity observed in Ref. [2]). This analysis also assumes kinesin tightly couples mechanical motion with chemical energy consumption [15] so that $W_{\text{chem}} = \langle v\rangle \Delta\mu_{\text{ATP}}/d$, for step size $d = 8$ nm and $\Delta\mu_{\text{ATP}} = 15\,k_B T$ [30, Chaps. 3 and 4].

For both myosin and kinesin, Fig. 5.6 shows that the predicted Pareto frontier (5.22) indeed bounds the experimentally observed performance. Consistently,

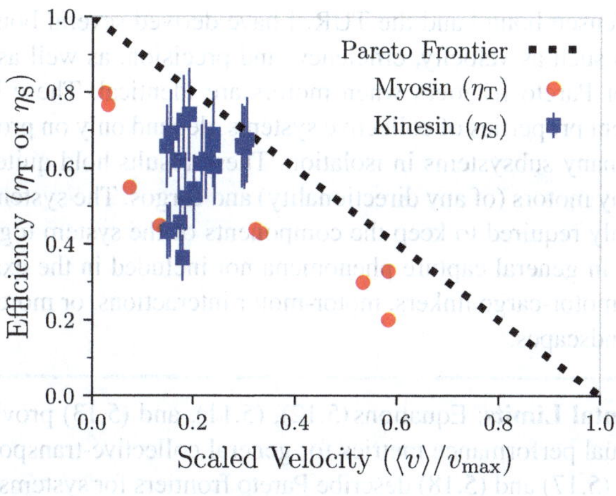

Fig. 5.6 Myosin motors across various animals, and kinesin motors pulling different-sized cargo, obey the theoretical Pareto frontier. Red points: experimental measurements of efficiency η_T and velocity $\langle v\rangle$ for myosin motors from different animal species [27]. Blue points: measured velocity and calculated Stokes efficiency for kinesin motors pulling diffusive cargo based on data from Ref. [2]. Dotted line: predicted Pareto frontier (5.22). Copyrighted by the American Physical Society [21]

theoretical studies of the trade-off between efficiency and velocity in other types of molecular machines have found that high velocity and high efficiency are mutually exclusive [31, 32].

5.8 Discussion

For motor-driven intracellular transport systems, the Jensen bound lower bounds the entropy production rate of a collective-transport system. This bound can be tighter or looser than the thermodynamic uncertainty relation (5.10), depending on the relative magnitudes of the bare collective diffusivity D_{bare} and the effective diffusivity D_{eff}. Due to its dependence solely on parameters and averaged quantities, the Jensen bound is much easier to compute than the TUR which depends on D_{eff} (a function of the variance, which requires more data to accurately estimate), provided that diffusion coefficients and driving forces are known for each subsystem in isolation.

Once these properties are known for a given set of subsystems, the Jensen bound is easily computed for any collective system assembled from a combination of such modular components. The TUR by contrast does not take advantage of information about the subsystems composing a collective system, and must be computed *de novo* for every such combination by measuring emergent properties of the collective system. This makes the Jensen bound particularly well suited for collective motor-driven transport systems, which are assembled out of parts (cargo and motors) that can be identified and studied in isolation.

Using the Jensen bound and the TUR, I have derived several bounds on performance metrics such as velocity, efficiency, and precision, as well as three analytic expressions for Pareto frontiers when motors are identical. These bounds, which restrict emergent properties of collective systems, depend only on properties of each of arbitrarily many subsystems in isolation. These results hold quite generally, for arbitrarily many motors (of any directionality) and cargos. The system's joint potential $V(\boldsymbol{x})$ is only required to keep the components of the system together at steady state, but may in general capture phenomena not included in the example, such as non-Hookean motor-cargo linkers, motor-motor interactions, or more complex periodic energy landscapes.

Fundamental Limits: Equations (5.12), (5.11), and (5.13) provide bounds on individual performance metrics for general collective-transport systems, while Eqs. (5.17) and (5.18) describe Pareto frontiers for systems with identical motors.

Numerical investigations show that the performance bounds and Pareto frontiers derived in this chapter are attainable for systems with no energy barriers. This is unsurprising, as it is well known that decreasing energy barriers (catalysis) speeds up a chemical reaction without affecting the energetics. All the bounds and frontiers are saturated for a model with only quadratic couplings between the cargo and each motor. This system, whose dynamics and thermodynamics have been solved analytically [8], is Pareto optimal for the class of systems considered here. More generally, the Jensen bound (5.7c) is always saturated for linear systems within the class of models considered here (Appendix. A.1 provides proof). The simulations focus on $N = 2$ motors due to computational constraints, however the derived bounds hold for arbitrarily large N; indeed, their utility is most significant for $N \gg 1$, where direct simulation is computationally intractable.

> **Design Principle**: Performance of a collective transport system can be maximized by making the overall potential energy landscape as close as possible to quadratic.

While the results of this chapter apply to a broad class of systems, they do rely on three key assumptions: (1) all components of the transport system stay together, achieving at long times the same mean velocity; (2) the dynamics are multipartite, such that the entropy production can be split into subsystem-specific contributions [18]; and (3) motor motion is tightly coupled to chemical-energy consumption. Multipartite dynamics are a standard assumption in stochastic thermodynamics [18, 33, 34], generally necessary to analyze the behaviour of multi-component systems. Experiments on kinesin [15, 16] and myosin [17] motors do support tight coupling between the mechanical and chemical degrees of freedom; nonetheless, futile cycles and backsteps have been observed to occur infrequently [35], and are beyond the scope of this analysis. I speculate that such phenomena can only degrade the performance metrics discussed, but generalizing these results to looser mechanochemical coupling will be an important future direction.

References

1. Klumpp S, Lipowsky R (2005) Cooperative cargo transport by several molecular motors. Proc Natl Acad Sci 102(48):17284–17289
2. Shtridelman Y, Cahyuti T, Townsend B, DeWitt D, Macosko JC (2008) Force-velocity curves of motor proteins cooperating in vivo. Cell Biochem Biophys 52(1):19–29
3. Shtridelman Y, Holzwarth GM, Bauer CT, Gassman NR, DeWitt DA, Macosko JC (2009) In vivo multimotor force-velocity curves by tracking and sizing sub-diffraction limited vesicles. Cell Mol Bioeng 2(2):190–199
4. Bhat D, Gopalakrishnan M (2016) Transport of organelles by elastically coupled motor proteins. Eur Phys J E 39(7):1–13
5. Bhat D, Gopalakrishnan M (2017) Stall force of a cargo driven by N interacting motor proteins. EPL (Eur Lett) 117(2):28004
6. Brown AI, Sivak DA (2019) Pulling cargo increases the precision of molecular motor progress. EPL (Eur Lett) 126(4):40004
7. McKinley SA, Athreya A, Fricks J, Kramer PR (2012) Asymptotic analysis of microtubule-based transport by multiple identical molecular motors. J Theor Biol 305:54–69

8. Leighton MP, Sivak DA (2022) Performance scaling and trade-offs for collective motor-driven transport. New J Phys 24(1):013009
9. Brown AI, Sivak DA (2019) Theory of nonequilibrium free energy transduction by molecular machines. Chem Rev 120(1):434–459
10. Seifert U (2012) Stochastic thermodynamics, fluctuation theorems and molecular machines. Rep Prog Phys 75(12):126001
11. Barato AC, Seifert U (2015) Thermodynamic uncertainty relation for biomolecular processes. Phys Rev Lett 114(15):158101
12. Gingrich TR, Horowitz JM, Perunov N, England JL (2016) Dissipation bounds all steady-state current fluctuations. Phys Rev Lett 116(12):120601
13. Horowitz JM, Gingrich TR (2020) Thermodynamic uncertainty relations constrain non-equilibrium fluctuations. Nat Phys 16(1):15–20
14. Pietzonka P, Barato AC, Seifert U (2016) Universal bound on the efficiency of molecular motors. J Stat Mech Theory Exp 12:124004
15. Schnitzer MJ, Block SM (1997) Kinesin hydrolyses one ATP per 8-nm step. Nature 388(6640):386–390
16. Visscher K, Schnitzer MJ, Block SM (1999) Single kinesin molecules studied with a molecular force clamp. Nature 400(6740):184–189
17. Toyoshima YY, Kron SJ, Spudich JA (1990) The myosin step size: measurement of the unit displacement per ATP hydrolyzed in an in vitro assay. Proc Natl Acad Sci 87(18):7130–7134
18. Horowitz JM (2015) Multipartite information flow for multiple Maxwell demons. J Stat Mech Theory Exp 2015(3):P03006
19. Reimann P, Van den Broeck C, Linke PH, Rubi JM, Pérez-Madrid A (2001) Giant acceleration of free diffusion by use of tilted periodic potentials. Phys Rev Lett 87(1):010602
20. Kawaguchi K, Uemura S, Ishiwata S (2003) Equilibrium and transition between single- and double-headed binding of kinesin as revealed by single-molecule mechanics. Biophys J 84(2):1103–1113
21. Leighton MP, Sivak DA (2022) Dynamic and thermodynamic bounds for collective motor-driven transport. Phys Rev Lett 129(11):118102
22. Kramers HA (1940) Brownian motion in a field of force and the diffusion model of chemical reactions. Physica 7(4):284–304
23. Vu HT, Chakrabarti S, Hinczewski M, Thirumalai D (2016) Discrete step sizes of molecular motors lead to bimodal non-Gaussian velocity distributions under force. Phys Rev Lett 117(7):078101
24. Evstigneev M, von Gehlen S, Reimann P (2009) Interaction-controlled Brownian motion in a tilted periodic potential. Phys Rev E 79(1):011116
25. Lathouwers E, Lucero JNE, Sivak DA (2020) Nonequilibrium energy transduction in stochastic strongly coupled rotary motors. J Phys Chem Lett 11(13):5273–5278
26. Müller P, Rogers KW, Yu SR, Brand M, Schier AF (2013) Morphogen transport. Development 140(8):1621–1638
27. Purcell TJ, Naber N, Franks-Skiba K, Dunn AR, Eldred CC, Berger CL, Málnási-Csizmadia A, Spudich JA, Swank DM, Pate E et al (2011) Nucleotide pocket thermodynamics measured by EPR reveal how energy partitioning relates myosin speed to efficiency. J Mol Biol 407(1):79–91
28. Piazzesi G, Lucii L, Lombardi V (2002) The size and the speed of the working stroke of muscle myosin and its dependence on the force. J Physiol 545(1):145–151
29. Einstein A et al (1905) On the motion of small particles suspended in liquids at rest required by the molecular-kinetic theory of heat. Annalen der physik 17(549–560):208
30. Milo R, Phillips R (2015) Cell biology by the numbers. Garland Science
31. Wagoner JA, Dill KA (2019) Opposing pressures of speed and efficiency guide the evolution of molecular machines. Mol Biol Evol 36(12):2813–2822
32. Wagoner JA, Dill KA (2021) Evolution of mechanical cooperativity among myosin II motors. Proc Natl Acad Sci 118(20)

33. Barato AC, Hartich D, Seifert U (2013) Information-theoretic versus thermodynamic entropy production in autonomous sensory networks. Phys Rev E 87(4):042104
34. Horowitz JM, Esposito M (2014) Thermodynamics with continuous information flow. Phys Rev X 4(3):031015
35. Clancy BE, Behnke-Parks WM, Andreasson JOL, Rosenfeld SS, Block SM (2011) A universal pathway for kinesin stepping. Nat Struct & Mol Biol 18(9):1020–1027

Chapter 6
Inferring Subsystem Efficiencies in Bipartite Molecular Machines

Molecular machines composed of coupled subsystems transduce free energy between different external reservoirs, in the process internally transducing energy and information. While subsystem efficiencies of these molecular machines have been measured in isolation, less is known about how they behave in their natural setting when coupled together and acting in concert. Here I derive upper and lower bounds on the subsystem efficiencies of a bipartite molecular machine. I demonstrate their utility by estimating the efficiencies of the F_o *and* F_1 *subunits of ATP synthase and that of kinesin pulling a diffusive cargo.*

6.1 Introduction

In addition to transducing free energy between different forms [1], a molecular machine can also transduce free energy within itself, between internally coupled components [2, 3]. F_oF_1−ATP synthase and transport motors like kinesin pulling diffusive cargo are two paradigmatic examples. Synthetic molecular machines are likewise often assembled from two or more components [4, 5]. To facilitate future design of synthetic molecular machines, it is critical to understand how components of a multicomponent molecular machine work both in isolation and together to transduce free energy.

Subsystems of bipartite molecular machines, such as the F_1 subunit [6], have been studied in isolation to determine their efficiency. Less, however, is known about how these subsystems perform when coupled together, as when performing their functions inside of biological organisms. For example, while experiments that measure motor efficiency typically apply a constant force, modelling efforts have shown that transport motors perform differently when pulling a diffusive cargo [7, 8]. Understanding molecular machines thus requires estimates of subsystem efficiencies within bipartite machines, in addition to their efficiencies in isolation.

M. Leighton, *Stochastic Thermodynamics of Multicomponent Molecular Machines*, Springer Theses, https://doi.org/10.1007/978-3-032-02204-2_6

In this chapter, I detail a new method to derive upper and lower bounds on the thermodynamic efficiencies of bipartite subsystems from any bounds on subsystem entropy production rates. As an example, I apply the recently proven Jensen lower bounds [9], which do not depend on detailed internal interactions, making them easy to compute even from limited data. I illustrate the utility of these bounds using experimental measurements to infer the efficiencies of F_o and F_1 when coupled together, as well as the efficiency of a kinesin motor while pulling a diffusive vesicular cargo. Ultimately this method allows for measurements of the efficiencies of subsystems in their natural settings, something inaccessible when studying them in isolation.

6.2 Bounds on Subsystem Efficiencies

The subsystem first (2.44a) and second laws (2.49) can be rearranged to provide two equalities for the transduced free energy,

$$\dot{\Sigma}_X - \beta \dot{W}_X = \beta \dot{E}_Y + \dot{I}_Y = \beta \dot{W}_Y - \dot{\Sigma}_Y. \tag{6.1}$$

Applying to Eq. (6.1) any lower bounds $\dot{\Sigma}_X^{\mathrm{LB}} \leq \dot{\Sigma}_X$ and $\dot{\Sigma}_Y^{\mathrm{LB}} \leq \dot{\Sigma}_Y$ on the subsystem entropy production rates yields upper and lower bounds on the transduced free energy:

$$\dot{\Sigma}_X^{\mathrm{LB}} - \beta \dot{W}_X \leq \beta \dot{E}_Y + \dot{I}_Y \leq \beta \dot{W}_Y - \dot{\Sigma}_Y^{\mathrm{LB}}. \tag{6.2}$$

Dividing Eq. (6.2) by $\dot{W}_Y$ yields upper and lower bounds on Y's efficiency:

$$\eta_{\mathrm{T}} \left(1 + \frac{\dot{\Sigma}_X^{\mathrm{LB}}}{-\beta \dot{W}_X}\right) \leq \eta_Y \leq 1 - \frac{\dot{\Sigma}_Y^{\mathrm{LB}}}{\beta \dot{W}_Y}. \tag{6.3}$$

Likewise, multiplying the reciprocal of Eq. 6.2 by $-\dot{W}_X$ yields upper and lower bounds on X's efficiency:

$$\eta_{\mathrm{T}} \left(1 - \frac{\dot{\Sigma}_Y^{\mathrm{LB}}}{\beta \dot{W}_Y}\right)^{-1} \leq \eta_X \leq \left(1 + \frac{\dot{\Sigma}_X^{\mathrm{LB}}}{-\beta \dot{W}_X}\right)^{-1}. \tag{6.4}$$

The two inequalities (6.3) and (6.4) constitute the most general form of the subsystem efficiency bounds, providing a recipe to derive bounds on subsystem efficiencies using the interactions of subsystems with their environments and lower bounds on their entropy production rates. These inequalities are valid for any lower bounds $\dot{\Sigma}_X^{\mathrm{LB}}$ and $\dot{\Sigma}_Y^{\mathrm{LB}}$, and are also valid for discrete degrees of freedom.

Inserting the subsystem second laws $\dot{\Sigma}_X^{\mathrm{LB}} = 0 = \dot{\Sigma}_Y^{\mathrm{LB}}$ into Eqs. (6.3) and (6.4) yields $\eta_{\mathrm{T}} \leq \eta_{X/Y} \leq 1$. Beyond the second law, however, the recently derived Jensen bound [9] gives tighter lower bounds for the overdamped bipartite Langevin dynamics considered here:

$$\beta\zeta_X\langle\dot{x}\rangle^2 \leq \dot{\Sigma}_X, \tag{6.5a}$$

$$\beta\zeta_Y\langle\dot{y}\rangle^2 \leq \dot{\Sigma}_Y. \tag{6.5b}$$

$\langle\dot{x}\rangle$ and $\langle\dot{y}\rangle$ are the steady-state average rates of change of the coordinates X and Y. Inserting these Jensen bounds into Eqs. (6.3) and (6.4) gives

$$\eta_\mathrm{T}\left(1+\frac{\zeta_X\langle\dot{x}\rangle^2}{-\dot{W}_X}\right) \leq \eta_Y \leq 1-\frac{\zeta_Y\langle\dot{y}\rangle^2}{\dot{W}_Y}, \tag{6.6a}$$

$$\eta_\mathrm{T}\left(1-\frac{\zeta_Y\langle\dot{y}\rangle^2}{\dot{W}_Y}\right)^{-1} \leq \eta_X \leq \left(1+\frac{\zeta_X\langle\dot{x}\rangle^2}{-\dot{W}_X}\right)^{-1}. \tag{6.6b}$$

This is a specific, immediately applicable version of Eqs. (6.3) and (6.4). Equations (6.6) bound internal energetic flows through subsystems, in terms of the experimentally accessible quantities ζ_X, ζ_Y, $\langle\dot{x}\rangle$, $\langle\dot{y}\rangle$, $\dot{W}_X$, and $\dot{W}_Y$ (recall that $\eta_\mathrm{T} = -\dot{W}_X/\dot{W}_Y$). These quantities solely depend on and characterize the interactions of the two subsystems with their environments; applying the bounds Eqs. (6.6a) and (6.6b) does not require any knowledge of the details of the coupling between subsystems.

Molecular machines that transduce free energy into directed motion rather than into stored free energy will often produce no output work ($\dot{W}_X = 0$, and thus also $\eta_X = 0 = \eta_\mathrm{T}$). It is then desirable to reformulate Eq. (6.6a) in a way that incorporates the Stokes efficiency and does not include division by $\dot{W}_X$. Substituting the definition of η_T, and identifying η_S (2.55), Eq. (6.6a) can be expressed as

$$\eta_\mathrm{T} + (1-\eta_\mathrm{T})\eta_\mathrm{S} \leq \eta_Y \leq 1-(1-\eta_\mathrm{T})\frac{\zeta_Y\langle\dot{y}\rangle}{\zeta_X\langle\dot{x}\rangle}\eta_\mathrm{S}. \tag{6.7}$$

When $\dot{W}_X = 0$ (and thus $\eta_\mathrm{T} = 0$), this bound significantly simplifies to

$$\eta_\mathrm{S} \leq \eta_Y \leq 1-\frac{\zeta_Y\langle\dot{y}\rangle}{\zeta_X\langle\dot{x}\rangle}\eta_\mathrm{S}. \tag{6.8}$$

6.3 Subsystem Efficiencies in ATP Synthase

I now apply Eqs. (6.6a) and (6.6b) to the molecular machine ATP synthase. The two coordinates Y and X correspond roughly to the rotational states of the c-ring inside the $\mathrm{F_o}$ subsystem and the γ-shaft inside the $\mathrm{F_1}$ subsystem, respectively [10].

Lacking experimental data on ATP synthesis and proton translocation rates, I assume that the $\mathrm{F_o}$ and $\mathrm{F_1}$ subsystems tightly couple [11] rotary motion with proton translocation [12] and ATP synthesis [6], respectively. The external work rates are then $\dot{W}_\mathrm{o} = \Delta\mu_{\mathrm{H}^+}\langle J_\mathrm{o}\rangle$ and $\dot{W}_1 = \Delta\mu_\mathrm{ATP}\langle J_1\rangle$, for the two subsystems' respective average rotation rates $\langle J_{\text{o}} \rangle$ and $\langle J_1 \rangle$ and chemical driving forces $\Delta\mu_{\text{H}^+}$ and $\Delta\mu_{\text{ATP}}$. This recasts the subsystem efficiency bounds (6.6a) and (6.6b) as

$$\eta_{\text{T}} \left(1 - \frac{\zeta_1 \langle J_1 \rangle}{\Delta\mu_{\text{ATP}}}\right) \leq \eta_{\text{o}} \leq 1 - \frac{\zeta_{\text{o}} \langle J_{\text{o}} \rangle}{\Delta\mu_{\text{H}^+}}, \tag{6.9a}$$

$$\eta_{\text{T}} \left(1 - \frac{\zeta_{\text{o}} \langle J_{\text{o}} \rangle}{\Delta\mu_{\text{H}^+}}\right)^{-1} \leq \eta_1 \leq \left(1 - \frac{\zeta_1 \langle J_1 \rangle}{\Delta\mu_{\text{ATP}}}\right)^{-1}. \tag{6.9b}$$

The six quantities composing the above bounds can be estimated from experimental data and theoretical calculations. Consider the bovine mitochondria, where many of the relevant quantities have been determined experimentally for ATP synthase far from stall. The chemical driving forces are estimated as $\Delta\mu_{\text{ATP}} \approx -7.5\, k_{\text{B}}T/\text{rad}$ and $\Delta\mu_{\text{H}^+} \approx 8.3\, k_{\text{B}}T/\text{rad}$ [13]. ATP can be synthesized at a rate of up to 440 molecules/second [14], and F_1 has been observed rotating at speeds of $\langle J_1 \rangle \approx 100\,\text{rot/s}$ [15]. Accordingly, I estimate the rotational flux to be $\langle J_1 \rangle \in [100, 150]\,\text{rot/s}$. Ref. [13] found $0.65 \leq \eta_{\text{T}} \leq 1$, so I take $\langle J_1 \rangle \leq \langle J_{\text{o}} \rangle \leq 1.4 \langle J_1 \rangle$. The friction coefficient of the γ-shaft rotating within the F_1 subsystem has been estimated to be of order $1.5 \times 10^{-2}\,\text{pN} \cdot \text{nm} \cdot \text{s/rad}^2$ [16]. Accordingly, I take $\zeta_1 \in [0.01, 0.03]\,\text{pN} \cdot \text{nm} \cdot \text{s/rad}^2$. Calculations of the rotational friction coefficients in Stokes flow [17] suggest $\zeta_{\text{o}} \approx \zeta_1/2$, so I take $\zeta_{\text{o}} \in [0.005, 0.015]\,\text{pN} \cdot \text{nm} \cdot \text{s/rad}^2$.

Figure 6.1 illustrates the joint range of subsystem efficiencies η_{o} and η_1 inferred from the bounds (6.9). This significantly constrains the possible subsystem efficiencies within ATP synthase, to $\eta_{\text{o}} \approx 0.5$–$0.85$ and $\eta_1 \approx 0.7$–0.85. Note that the size and location of the inferred region are somewhat sensitive to the parameter estimates; more precise measurements of physical parameters would allow for tighter thermodynamic inference. Because of the functional form of the Jensen bound, the inferred region is also smaller for higher friction and higher coordinate rates of change.

6.4 Efficiency of a Transport Motor Pulling a Diffusive Cargo

Taking Y and X to be the respective one-dimensional positions along a microtubule of a transport motor and its cargo, Eq. (6.8) allows estimation of the motor efficiency η_{M}, quantifying the free-energetic efficiency of a motor pulling against the fluctuating force arising from the motion of the diffusive cargo. Since a coupled motor and cargo have equal average velocity [9], $\langle \dot{x} \rangle = \langle \dot{y} \rangle = \langle v \rangle$, Eq. (6.8) further simplifies to

$$\eta_{\text{S}} \leq \eta_{\text{M}} \leq 1 - \frac{D_{\text{c}}}{D_{\text{m}}} \eta_{\text{S}}. \tag{6.10}$$

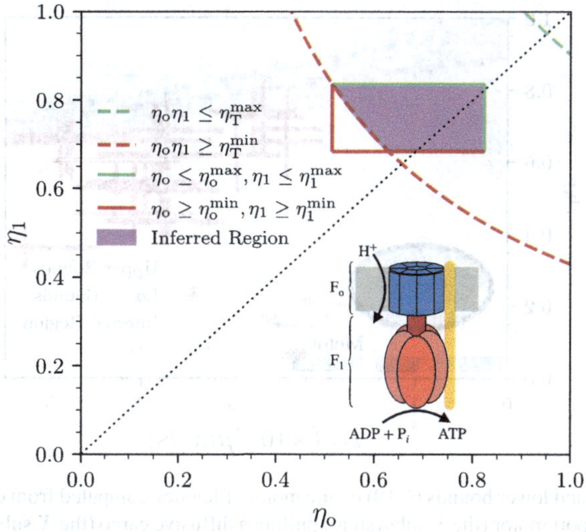

Fig. 6.1 Inferred subsystem efficiencies η_o for F_o (the *Y* subsystem) and η_1 for F_1 (the *X* subsystem) in F_oF_1−ATP synthase. Red horizontal and vertical lines: minimum possible values of the lower bounds (6.9) on η_1 and η_o, respectively, that are consistent with the estimated parameter ranges detailed in the main text. Green horizontal and vertical lines: maximum possible values of the two upper bounds (6.9). Red and green dashed curves: minimum and maximum possible values of the product $\eta_o \cdot \eta_1 = \eta_T$ given my parameter estimates. Purple region: efficiencies satisfying all bounds. Copyrighted by the American Physical Society [18]

Here the friction coefficients of Eq. (6.8) have been replaced with diffusion coefficients (using the fluctuation-dissipation relation [19]) which are more natural for the motor-cargo system.

Estimating η_M requires measurements of the diffusion coefficients D_c and D_m, the average velocity $\langle v \rangle$, and the chemical power consumption $\dot{W}_M$ by the motor. Reference [20] provides experimental measurements of average velocity for single transport motors (motor number inferred from multimodal velocity distributions) pulling vesicles, as a function of the vesicle diameter. Cargo diffusivity D_c is estimated from the measured diameter using the Stokes-Einstein relation [21] and reported measurements of temperature and viscosity. I estimate D_m by fitting the resulting $\langle v \rangle$ as a function of D_c to the theoretical prediction $\langle v \rangle / v_{max} = (1 + D_m/D_c)^{-1}$ [22] with $v_{max} = 2\,\mu\text{m/s}$ (the maximum velocity observed in Ref. [20]). Finally, I assume the transport motor tightly couples mechanical motion with chemical energy consumption [23] so that $\dot{W}_M = \langle v \rangle \Delta\mu_{ATP}/d$, for step size $d = 8$ nm and $\Delta\mu_{ATP} = 15\,k_BT$ [24, Chaps. 3 and 4].

Figure 6.2 shows the upper and lower bounds on η_M (6.10) inferred from the above estimates and experimental data. This method for the first time significantly constrains η_M, suggesting it decreases from ~0.85 to ~0.75 as D_c increases from $2\,\mu\text{m}^2/\text{s}$ to $5.5\,\mu\text{m}^2/\text{s}$ (corresponding to vesicle diameters from $1.1\,\mu\text{m}$ to $0.4\,\mu\text{m}$).

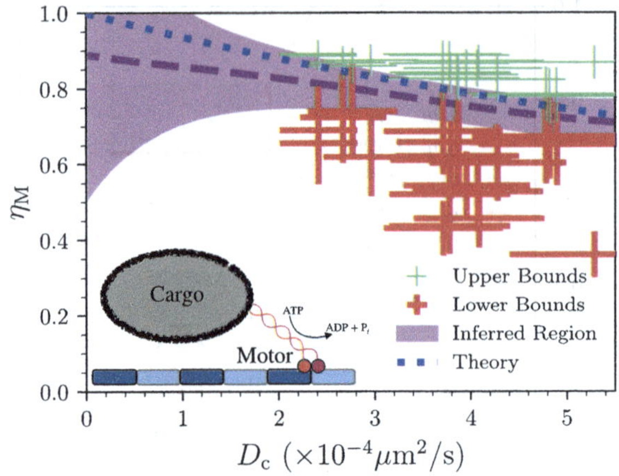

Fig. 6.2 Upper and lower bounds (6.10) on the motor efficiency computed from experimental data for a single transport motor (the Y subsystem) pulling a diffusive cargo (the X subsystem) [20]. For each paired velocity and cargo diameter from Ref. [20], red and green points respectively denote lower and upper bounds (Eq. (6.10)) on the motor efficiency, with error bars showing experimental error propagated through Eq. (6.10). Purple dashed curve: nonlinear best-fit curve separating the upper and lower bounds computed using a nonlinear support vector machine [25]. Purple shaded region: corresponding confidence interval. Blue dot-dashed line: theoretical prediction from Chap. 4. Copyrighted by the American Physical Society [18]

These estimates are consistent with my earlier theoretical prediction from Chap. 4, $\eta_M = (1 + D_c/D_m)^{-1}$, which falls entirely within the inferred region of Fig. 6.2.

While the cargo does not have a thermodynamic efficiency since it does not provide output work, it is nonetheless possible to define an efficiency metric for the cargo subsystem. Similar to the definition of the Stokes efficiency in Chap. 2, considering the output of the cargo as work against a viscous friction force suggests the definition of the *Stokes subsystem efficiency*:

$$\eta_C \equiv \frac{\zeta_c \langle v \rangle^2}{\dot{E}_M + k_B T \dot{I}_M}. \tag{6.11}$$

η_C quantifies the efficiency at which the cargo dynamics transduce free energy from the motor into work against viscous friction.

Applying the Jensen bound to the subsystem second law for the cargo yields the upper bound $\eta_C \leq 1$, making this a well-defined efficiency metric. Similarly, applying the Jensen bound to the motor second law yields a lower bound on η_C. Thus like the other subsystem efficiencies considered in this chapter, the cargo Stokes subsystem efficiency is constrained by upper and lower bounds:

$$\eta_S \left(1 - \frac{\langle v \rangle d}{D_m \beta \Delta \mu_{ATP}}\right)^{-1} \leq \eta_C \leq 1. \tag{6.12}$$

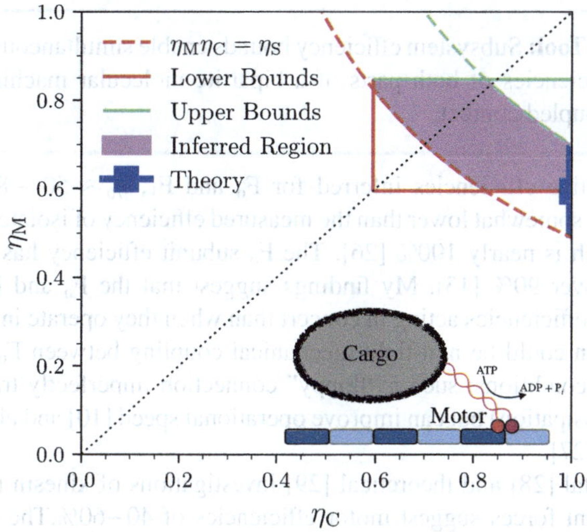

Fig. 6.3 Inferred subsystem efficiencies η_M for kinesin (the Y subsystem) and η_C for its diffusive cargo (the X subsystem). Red horizontal and vertical lines: minimum possible values of the lower bounds (6.10) and (6.12) on η_M and η_C, respectively, that are consistent with the estimated parameters detailed in the main text. Green horizontal and vertical lines: maximum possible values of the two upper bounds (6.10) and (6.12). Red and green dashed curves: minimum and maximum possible values of the product $\eta_M \cdot \eta_C = \eta_S$ given parameter estimates. Purple region: efficiencies satisfying all bounds. Blue point: theoretical prediction from Chap. 4. Copyrighted by the American Physical Society [18]

As with F_oF_1-ATP synthase, it is now possible to plot simultaneous constraints on the motor and cargo subsystem efficiencies, which I do in Fig. 6.3, again based on experimental data from Ref. [20]. Here I take averages over all data points shown in Fig. 6.2 to obtain estimates for D_c and $\langle v \rangle$.

6.5 Discussion

I derived general lower and upper bounds on the efficiencies of two subsystems composing a bipartite molecular machine in their natural setting, as opposed to in isolation as in typical single-molecule experiments. The measurable quantities required to compute the bounds depend only on the interactions of each molecular machine with its environment; details of the coupling between subsystems need not be understood. Quantifying subsystem efficiencies allows us to determine where free energy is lost in multi-component systems, which will ultimately be critical for the future engineering of synthetic molecular machines.

Inference Tool: Subsystem efficiency bounds enable simultaneous inference of the efficiencies of both parts of a bipartite molecular machine in their natural, coupled context.

The respective efficiencies inferred for F_o and F_1, $\eta_o \approx 50 - 85\%$ and $\eta_1 \approx 70 - 85\%$, are somewhat lower than the measured efficiency of isolated F_1 hydrolyzing ATP, which is nearly 100% [26]. The F_o subunit efficiency has likewise been estimated at over 90% [13]. My findings suggest that the F_o and F_1 subsystems have different efficiencies acting in concert than when they operate in isolation. One possible reason could be non-tight mechanical coupling between F_o and F_1 under physiological conditions; such a "floppy" connection imperfectly transfers energy (increasing dissipation) but can improve operational speed [10] and allows for information flows [27].

Experimental [28] and theoretical [29] investigations of kinesin motors pulling against constant forces suggest motor efficiencies of 40–60%.The inferred range of 70–90% is slightly higher, suggesting that transport motors may attain higher efficiencies when pulling against the variable load produced by a diffusive cargo, their main function within cells.

It is important to note conceptual differences between previous subsystem efficiencies and those inferred in this chapter. Conventional single-molecule experiments measuring subsystem efficiency (such as Ref. [26]) typically consider transduction to and from deterministic external reservoirs, hence preclude information flows and are limited to work. The subsystem efficiencies (Eqs. (2.54a) and (2.54b)) consider transduction to and from a strongly coupled stochastic subsystem, naturally including information transmission and hence encompassing all free-energy transduction. In contrast with conventional experimental efficiencies that study isolated subsystems in artificial environments, the subsystem efficiencies describe their behaviour in their natural, coupled, *in vivo* context.

The main results are derived here for X and Y subsystems fully characterized by one-dimensional degrees of freedom, but are more general. The x and y coordinates will in most cases be coarse-grained over many internal degrees of freedom; such a coarse-graining underestimates the true entropy production [30, 31], so the efficiency bounds would loosen but remain valid because they derive from lower bounds on the entropy production rates.

Finally, while I employed the Jensen bound (6.5) to derive lower bounds for $\dot{\Sigma}_X$ and $\dot{\Sigma}_Y$ and thus derive Eqs. (6.6a) and (6.6b), this framework is far more general. Any set of lower or upper bounds on $\dot{\Sigma}_X$ and $\dot{\Sigma}_Y$ could be inserted into Eq. (6.1) to obtain different subsystem efficiency bounds. Promising examples include the recently derived bipartite TUR [32].

References

1. Brown AI, Sivak DA (2019) Theory of nonequilibrium free energy transduction by molecular machines. Chem Rev 120(1):434–459
2. Large SJ, Ehrich J, Sivak DA (2021) Free-energy transduction within autonomous systems. Phys Rev E 103(2):022140
3. McGrath T, Jones NS, Ten Wolde PR, Ouldridge TE (2017) Biochemical machines for the interconversion of mutual information and work. Phys Rev Lett 118(2):028101
4. Wilson MR, Solà J, Carlone A, Goldup SM, Lebrasseur N, Leigh DA (2016) An autonomous chemically fuelled small-molecule motor. Nature 534(7606):235–240
5. Courbet A, Hansen J, Hsia Y, Bethel N, Park Y-J, Xu C, Moyer A, Boyken SE, Ueda G, Nattermann U et al (2022) Computational design of mechanically coupled axle-rotor protein assemblies. Science 376(6591):383–390
6. Shoichi T, Takahiro W-N, Tetsuaki O, Seishi K, Eiro M (2011) Thermodynamic efficiency and mechanochemical coupling of F_1-ATPase. Proc Natl Acad Sci 108(44):17951–17956
7. Eva Z, Udo S (2015) Effective rates from thermodynamically consistent coarse-graining of models for molecular motors with probe particles. Phys Rev E 91(2):022709
8. Brown AI, Sivak DA (2019) Pulling cargo increases the precision of molecular motor progress. EPL (Europhys Lett) 126(4):40004
9. Leighton MP, Sivak DA (2022) Dynamic and thermodynamic bounds for collective motor-driven transport. Phys Rev Lett 129(11):118102
10. Lathouwers E, Lucero JNE, Sivak DA (2020) Nonequilibrium energy transduction in stochastic strongly coupled rotary motors. J Phys Chem Lett 11(13):5273–5278
11. Soga N, Kimura K, Kinosita K, Yoshida M, Suzuki T (2017) Perfect chemomechanical coupling of F_oF_1-ATP synthase. Proc Natl Acad Sci 114(19):4960–4965
12. Marciniak A, Chodnicki P, Hossain KA, Slabonska J, Czub J (2022) Determinants of directionality and efficiency of the ATP synthase F_o motor at atomic resolution. J Phys Chem Lett 13:387–392
13. Silverstein TP (2014) An exploration of how the thermodynamic efficiency of bioenergetic membrane systems varies with c-subunit stoichiometry of F_oF_1 ATP synthases. J Bioenerg Biomembr 46(3):229–241
14. Akemi M-Y, Youssef H (1988) Estimation of the turnover number of bovine heart F_oF_1 complexes for ATP synthesis. Biochemistry 27(1):335–340
15. Carsten E, Gabriele D-H, Karlheinz A (1997) Turnover number of Escherichia coli F_oF_1 ATP synthase for ATP synthesis in membrane vesicles. Eur J Biochem 243(1–2):336–343
16. Buzhynskyy N, Sens P, Prima V, Sturgis JN, Scheuring S (2007) Rows of ATP synthase dimers in native mitochondrial inner membranes. Biophys J 93(8):2870–2876
17. Kei-ichi O, Gerhard H (2015) Elasticity, friction, and pathway of γ-subunit rotation in F_oF_1-ATP synthase. Proc Natl Acad Sci 112(34):10720–10725
18. Leighton MP, Sivak DA (2023) Inferring subsystem efficiencies in bipartite molecular machines. Phys Rev Lett 130(17):178401
19. Udo S (2012) Stochastic thermodynamics, fluctuation theorems and molecular machines. Rep Prog Phys 75(12):126001
20. Shtridelman Y, Cahyuti T, Townsend B, DeWitt D, Macosko JC (2008) Force-velocity curves of motor proteins cooperating in vivo. Cell Biochem Biophys 52(1):19–29
21. Albert E et al (1905) On the motion of small particles suspended in liquids at rest required by the molecular-kinetic theory of heat. Ann Phys 17(549–560):208
22. Leighton MP, Sivak DA (2022) Performance scaling and trade-offs for collective motor-driven transport. New J Phys 24(1):013009
23. Schnitzer MJ, Block SM (1997) Kinesin hydrolyses one ATP per 8-nm step. Nature 388(6640):386–390
24. Milo R, Phillips R (2015) Cell biology by the numbers. Garland Science

25. Suykens JAK (2001) Nonlinear modelling and support vector machines. In: IMTC 2001. Proceedings of the 18th IEEE Instrumentation and Measurement Technology Conference. Rediscovering Measurement in the Age of Informatics (Cat. No. 01CH 37188), vol 1, pp 287–294. IEEE
26. Shoichi T, Tetsuaki O, Takahiro W-N, Hiroshi T, Seishi K, Eiro M (2010) Nonequilibrium energetics of a single F_1-ATPase molecule. Phys Rev Lett 104(19):198103
27. Lathouwers E, Sivak DA (2022) Internal energy and information flows mediate input and output power in bipartite molecular machines. Phys Rev E 105(2):024136
28. Jonathon H (1996) The movement of kinesin along microtubules. Annu Rev Physiol 58(1):703–729
29. Lau AWC, Lacoste D, Mallick K (2007) Nonequilibrium fluctuations and mechanochemical couplings of a molecular motor. Phys Rev Lett 99(15):158102
30. Massimiliano E (2012) Stochastic thermodynamics under coarse graining. Phys Rev E 85(4):041125
31. Horowitz JM (2015) Multipartite information flow for multiple Maxwell demons. J Stat Mechan Theory Exper 2015(3):P03006
32. Tomohiro T, Van Tan V, Keiji S (2023) Universal bounds on the performance of information-thermodynamic engine. Phys Rev Res 5(4):043280

Chapter 7
Information Arbitrage in Bipartite Heat Engines

Heat engines and information engines have each historically served as motivating examples for the development of thermodynamics. While these two types of systems are typically thought of as two separate kinds of machines, recent empirical studies of specific systems have hinted at possible connections between the two. Inspired by molecular machines in the cellular environment, which in many cases have separate components in contact with distinct sources of fluctuations, we study bipartite heat engines. We show that a bipartite heat engine can only produce net output work by acting as an information engine. Conversely, information engines can only extract more work than the work consumed to power them if they have access to different sources of fluctuations, i.e., act as heat engines. We illustrate these findings first through an analogy to economics and a cyclically controlled 2D ideal gas. We then explore two analytically tractable model systems in more detail: a Brownian-gyrator heat engine which we show can be reinterpreted as a feedback-cooling information engine, and a quantum-dot information engine which can be reinterpreted as a thermoelectric heat engine. These results suggest design principles for both heat engines and information engines at the nanoscale, and ultimately imply constraints on how free-energy transduction is carried out in biological molecular machines.

7.1 Introduction

In contrast to the classical thermodynamic arena of heat engines operating between different temperatures, biological processes are typically assumed to be isothermal, with fluctuations often treated as homogeneous and isotropic. However, recent experimental and theoretical developments have shed light on possible departures from uniform fluctuations. For example, experiments suggest the mitochondrial temperature could be as much as 10K hotter than the rest of the cell [1–4]. This temperature difference could conceivably be accessed by the molecular machine ATP synthase

M. Leighton, *Stochastic Thermodynamics of Multicomponent Molecular Machines*, Springer Theses, https://doi.org/10.1007/978-3-032-02204-2_7

which straddles the mitochondrial membrane. As another example, light-harvesting machines like photosystem II [5] are driven out of equilibrium by solar photons. These light-induced reactions can be treated as coupling to a heat bath at the temperature of the photon source [6–8]. Lastly, the cellular interior supports a host of *active fluctuations* [9–12] powered by metabolic activity via the motion of large cytosolic components, for example enzymes and related complexes [13, 14] or the cytoskeletal network [15].

Such biological systems are typically composed of interacting degrees of freedom, which may thus be separately influenced by fluctuations of different strengths. The theory of bipartite stochastic thermodynamics [16–18] describes energy and entropy balance at the level of individual subsystems, and allows for quantification of internal energy and information flows [19] between coupled subsystems. In such setups, different sources of fluctuations can be leveraged to improve performance. For example, active fluctuations speed *in vitro* kinesin operation [20] and enzymatic catalysis [21], and temperature differences increase output work in a model for ATP synthase [22]. These effects are reminiscent of classical heat engines that alternately couple to different reservoirs at distinct temperatures. However, microscopic biological systems differ in that they are composed of interacting subsystems which each experience fluctuations from different sources. This motivates the study of two-component heat engines in which each part interacts with a heat bath at a different temperature. We call such systems *bipartite heat engines*.

Nanoscale stochastic systems can also use information as a thermodynamic resource to behave as information engines. Information engines fundamentally require a setup with two components: a *controller* and a *controlled system*. Information can then serve as a thermodynamic resource to make the controlled system, when considered on its own, do something seemingly forbidden by the second law, e.g., convert heat entirely into work. This comes at the controller's expense because the apparent second-law violation entails an energetic cost [23]—through Landauer's principle [24]—for performing feedback control and erasing previously acquired information.

Because of their interacting components and the relevance of (thermal) fluctuations, it is natural to ask whether molecular machines behave as information engines. Assuming a bipartite setup permits quantification of the information thermodynamics of such systems analogously to their energetics [25], by calculating an *information flow* [19, 26, 27] one can quantify the extent to which information is transduced through a composite system's dynamics. This setup has been used to bound the dissipation of molecular sensors [28–30] and study the role information plays in bipartite molecular machines [22, 31–35].

It has recently been suggested that information engines designed to leverage nonequilibrium fluctuations [36–38] can greatly outperform their purely thermally driven counterparts. Output power can even surpass minimum control costs, rendering the information engine an energy harvester that operates between two reservoirs, the equilibrium thermal fluctuations affecting the controller and the nonequilibrium fluctuations affecting the controlled system. This setup is suggestively similar to that of a bipartite heat engine with a "cold" and a "hot" subsystem. In fact, like heat engines, such information engines are constrained by the Carnot bound, hinting at connections between the two engine types [39–42].

This chapter elucidates a connection between heat engines exploiting the flow of heat from hot to cold, and information engines implementing Maxwell-demon-like exploitation of fluctuations. To broaden the perspective, I illustrate that bipartite heat engines, where two engine components are each in contact with distinct reservoirs at different temperatures, can indeed be understood as information engines, and vice versa. To operate as a heat engine and extract energy from the temperature difference, the two components need to work together by exchanging entropy—in the form of information. This cooperation can be understood as an information engine in which the "colder" component acts as a controller that exploits the fluctuations of the "hotter" component. I illustrate these core findings using an analogy from economics, followed by a simple example consisting of a Carnot cycle on a 2D ideal gas with anisotropic temperature which is revealed to be a disguised information engine.

Building on the theory of bipartite stochastic thermodynamics, I derive the information flow arbitrage relation (Eq. (7.14)): an inequality bounding the output work of a bipartite heat engine by the product of the temperature difference between the two reservoirs and an internal information flow between the two subsystems. This result shows that a bipartite heat engine can only achieve net output work by supporting an information flow between its two subsystems, thus acting as an information engine. Conversely, a bipartite information engine can only produce more work than the energy cost required to run the controller when it operates between two temperatures, thus acting as a heat engine. This shift in perspective helps to establish the information engine as a useful mechanism for work extraction, and implies that any bipartite heat engine must implicitly contain this information-engine mechanism: Maxwell's demon lies at the heart of many real-world heat engines.

7.2 Heat Engines are Entropy Arbitrageurs

Fundamentally, heat engines trade energy with heat reservoirs. They receive a certain amount of energy in the form of heat from the hot reservoir, and give a lesser amount of heat to the cold reservoir, withdrawing the energy difference as output work. Since energy is conserved, the work extracted is equal to the difference in heats.

Flows of another quantity, *entropy*, determine how input heat can be split among output work and output heat. The input heat from the hot reservoir comes with an increase in the entropy of the heat engine, with the ratio of energy to entropy bounded by the temperature of the reservoir (multiplied by Boltzmann's constant k_B). Because of the lower temperature of the cold reservoir, getting rid of the entropy only requires giving off a smaller amount of heat.

The greater the temperature difference, the less of the input energy needs to flow to the cold reservoir to maintain entropy balance, and thus the more energy can be extracted. Figure 7.1a illustrates the energy and entropy flows in a heat engine in

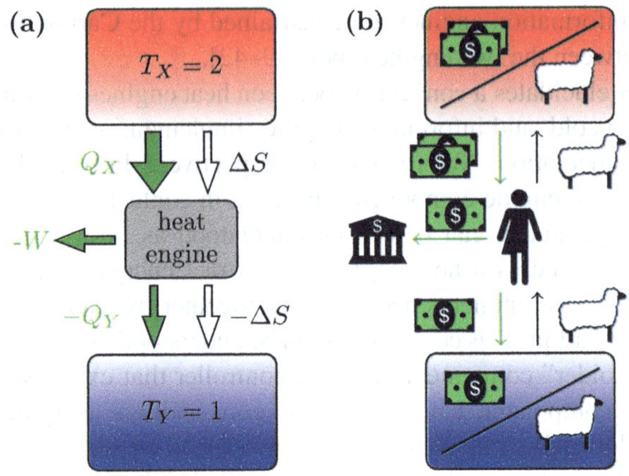

Fig. 7.1 Heat engine as entropy arbitrageur. **a** Energy (green filled arrows) and entropy (unfilled arrows) flows in an ideal heat engine operating between a hot ($T_X = 2$) and a cold ($T_Y = 1$) reservoir. **b** Arbitrage analogy from economics. The heat engine is an arbitrageur buying sheep (entropy reduction) for a low price and selling sheep (entropy increase) for a higher price at a different market, pocketing the difference in money (energy). Copyrighted by the American Physical Society [43]

contact with two heat reservoirs at temperature $T_X = 2$ and $T_Y = 1$, respectively. For simplicity, we stick to a temperature ratio of $T_X/T_Y = 2$ throughout this chapter. We take the convention that work and heat flows into the system are positive.

The analysis invites an analogy from economics: When in two markets the prices for the same good differ, a market participant can make risk-free profit by buying from one market at a lower price and selling at another market for a higher price, pocketing the difference. This practice is *arbitrage* [44] and people engaging in it are *arbitrageurs*. The heat engine described above is such an arbitrageur: It "buys" a reduction in entropy from the cold reservoir for a smaller amount of energy and "sells" an equal entropy increase to the hot reservoir for a larger amount of energy, pocketing the difference. Inspired by [45], we depict the analogy with an arbitrageur trading sheep at different markets in Fig. 7.1b. In this analogy money corresponds to energy, while sheep correspond to entropy reduction—since by the second law entropy cannot spontaneously decrease, decreasing entropy most cost something.

Trading sheep this way drives prices in the two markets, through the forces of supply and demand, toward equality, known as an *arbitrage equilibrium*; aptly named, since in the thermodynamic context the heat flow through the engine eventually leads to an intermediate temperature in the reservoirs and a global *thermodynamic equilibrium*. Just as processes in nature evolve towards an equilibrium, so do economic forces impel market participants to engage in arbitrage, pushing the market towards equilibrium. This analogy between heat engines and market arbitrage is not new [46, 47] and, though different from the approach here, has even been used in the context of an information engine [48].

We consider the analogy particularly illuminating because in the same way in which a trader cannot increase the number of sheep by making them out of thin air (except by spending money) the second law forbids decreasing entropy (except by spending energy to do so). Hence, in both cases optimal efficiency is achieved by conserving the traded good. Inefficiencies in thermodynamic engines are expressed by unnecessary entropy production that increases the amount of heat that needs to be dissipated to break even. Similarly, if the arbitrageur somehow loses some sheep between the markets without compensation, they sell fewer sheep and make a smaller profit.

7.2.1 Bipartite Heat Engines: Information Arbitrage

Not all engines are in simultaneous or alternating contact—as in conventional Carnot analysis—with two heat reservoirs. Instead, some engines are composed of distinct *subsystems* that are each in permanent contact with a different reservoir at a given temperature. Macroscopic examples include thermoelectric devices, where two coupled junctions in contact with different temperatures collectively achieve work output [49]. Microscopic examples include the molecular machines mentioned in the Introduction, which are made up of different components that can be exposed to different sources of fluctuations from temperature gradients [1–4], hot thermal radiation [8], or active fluctuations [9, 10, 13–15, 20].

Motivated by these examples, I now consider *bipartite heat engines*, where two subsystems each interact with only one reservoir at a distinct temperature. Collectively, the subsystems can act as a heat engine, conducting heat from hot to cold and producing work output. Figure 7.2a depicts such a setup: The larger input heat Q_X is entirely converted to output work $-W_X$, while the smaller input work W_Y is entirely converted to output heat $-Q_Y$. Comparing with Fig. 7.1a, $-W = -W_X - W_Y > 0$, i.e., net output work is positive.

The setup portrayed in Fig. 7.2a does not require a flow of energy through the machine because heat and work are converted locally in each subsystem; however, the machine requires a flow of entropy from the hot subsystem to the cold subsystem. In stochastic thermodynamics, entropy transduction between components is called *information flow* [18, 19, 26]. Section 7.3 gives a precise mathematical statement of information flow, but for now we make do with an intuitive explanation: In information theory, mutual information measures the mutual dependence between two variables. Information flow measures how much each subsystem tends to increase or decrease the mutual information between their statistical states. By the equivalence of information-theoretic (Shannon) entropy and thermodynamic entropy, subsystems may exchange thermodynamic entropy with each other by altering their mutual information.

This flow of information is the hallmark of an *information engine* or Maxwell demon, which achieves conversion of heat to work using information, in apparent violation of the second law. In this setup we immediately see that the apparent paradox

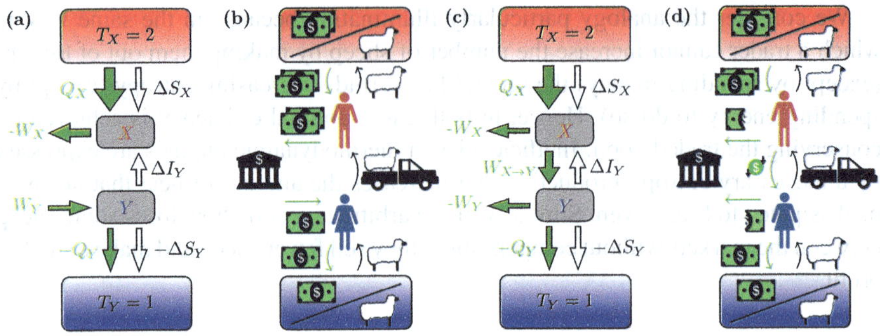

Fig. 7.2 Ideal bipartite heat engines and their economic analogs. **a** Energy (green filled arrows) and entropy (unfilled arrows) flows. The flow of entropy between engine components X and Y is conventionally identified as an information flow (in the opposite direction). In **a**, X and Y cannot exchange energy, so Y requires external input work. **b** Arbitrage analogy of (**a**): Two market participants must work together to realize the arbitrage scheme, requiring a flow of sheep from the lower-priced to the higher-priced market as well as a flow of money from the higher-priced market to the lower-priced market. **c** Same as **a** but with X and Y each extracting work. **d** Arbitrage analogy of (**c**). Copyrighted by the American Physical Society [43]

results from only focusing on the X-subsystem in Fig. 7.2a, which indeed transforms input heat to output work, and ignoring the Y-subsystem, which dissipates input work as output heat.

We can also understand the necessity of a flow of entropy in terms of the previous arbitrage analogy, Fig. 7.2b: Because each trader only has access to one market, they need to work together to make a net profit. Since they cannot produce new sheep (no decreasing entropy), a necessary requirement for their arbitrage scheme to work is that sheep are transported from the market with the lower exchange rate to that with the higher exchange rate.

Finally, bipartite heat engines may differ in which subsystem is capable of extracting work. For example, if the Y-subsystem has no access to a work source, the machine must transduce work from the hot to the cold side to "pay for" reducing its entropy by releasing heat to the cold reservoir. In terms of the analogy, in this case money must be transferred from one market participant to the other. Of course, multiple different schemes of money transfer could be set up. For example, the two arbitrageurs could equally share profits, as illustrated in Fig. 7.2c and d. Notice, however, that regardless of the details of the energy (money) extraction, entropy (sheep) must be transported from hot to cold reservoirs (lower- to higher-exchange-rate markets).

7.2.2 Example: Carnot Cycle for 2D Ideal Gas

If bipartite heat engines necessitate an information flow to generate net output power, one should be able to interpret a given bipartite heat engine in terms of an equivalent information engine. This section illustrates this principle in a slight

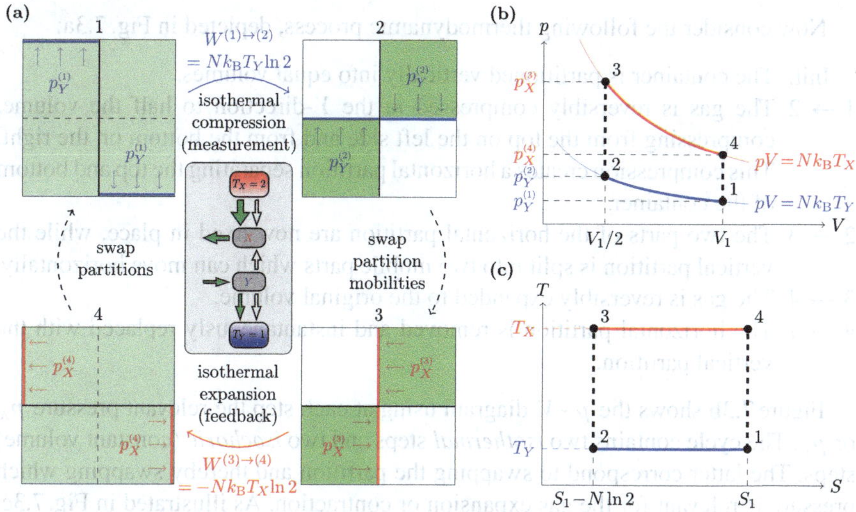

Fig. 7.3 Heat engine using a two-dimensional ideal gas in contact with heat baths acting in different directions. **a** Illustration of thermodynamic cycle. Small arrows indicate respective pressures in X and Y directions. Central inset: thermodynamic diagram showing presence/absence and directionalities of all energy, entropy, and information flows over the course of one cycle. **b** $p - V$ diagram indicating the pressures on the mobile partitions (p) and volumes (V) for each step in the cycle. (c) Corresponding $T - S$ diagram. In **b** and **c**, red and blue curves denote isotherms at T_X and T_Y, respectively, while black dashed lines denote isochores (or equivalently adiabats). Copyrighted by the American Physical Society [43]

modification of a well-known thermodynamic process. Consider a 2D ideal gas, depicted in Fig. 7.3a, in which the position and momentum coordinates in the vertical direction are assumed to not interact with those in the horizontal direction, and the coordinates in the two directions are in contact with different thermal reservoirs at respective temperatures T_X and $T_Y < T_X$. This idealized setup could be achieved, for example, with a monatomic gas cooled to low temperature in a container with perfectly elastic walls held at different temperatures. In the following, we show how such a setup is harnessed to execute the thermodynamic cycle of a heat engine. Then, in Sect. 7.2.3 we interpret the thermodynamic process in terms of the well-known Szilard engine. Similar systems have been proposed and analyzed previously [41, 42, 50], but without considering connections between the two engine types.

The gas is confined to a square container of side length L and exerts different pressures p_X and p_Y on the vertical and horizontal container walls, respectively. The ideal-gas law then gives relationships between the pressures p_X and p_Y (defined as forces per unit length), container volume $V = L^2$, and temperatures:

$$p_X V = Nk_B T_X \tag{7.1a}$$

$$p_Y V = Nk_B T_Y \,, \tag{7.1b}$$

where N is the number of molecules.

Now consider the following thermodynamic process, depicted in Fig. 7.3a:

Init. The container is partitioned vertically into equal volumes.

$1 \to 2$ The gas is reversibly compressed in the Y-direction to half the volume, compressing from the top on the left side and from the bottom on the right. This compression creates a horizontal partition separating the top and bottom of the container.

$2 \to 3$ The two parts of the horizontal partition are now fixed in place, while the vertical partition is split into two mobile parts which can move horizontally.

$3 \to 4$ The gas is reversibly expanded to the original volume.

$4 \to 1$ The horizontal partition is removed and instantaneously replaced with the vertical partition.

Figure 7.3b shows the p–V diagram using at each step the relevant pressure p_X or p_Y. The cycle contains two *isothermal* steps and two *isochoric* (constant volume) steps. The latter correspond to swapping the partition and thereby swapping which pressure is relevant for the gas expansion or contraction. As illustrated in Fig. 7.3c, this instantaneous swapping of the container partitions is not only isochoric but also *adiabatic* since no heat is exchanged with the heat baths. This is possible because the engine essentially has two working media at different fixed temperatures: the X and Y components of the ideal gas.

During the isochoric steps ($2 \to 3$ and $4 \to 1$), no heat is exchanged with the baths, and the compression ($1 \to 2$) and expansion ($3 \to 4$) steps are isothermal at respective temperatures T_Y and T_X. Hence the work done on the gas during compression is $W_{1\to2} = Nk_B T_Y \ln 2$ and during expansion is $W_{3\to4} = -Nk_B T_X \ln 2$. Heat flows into the gas during the expansion, $Q_{3\to4} = Nk_B T_X \ln 2$, and out of the gas during compression, $Q_{1\to2} = -Nk_B T_Y \ln 2$. Hence, in one cycle the heat engine outputs net work

$$-W_{1\to2} - W_{3\to4} = Nk_B (T_X - T_Y) \ln 2 \tag{7.2}$$

at Carnot efficiency

$$\eta = \frac{-W_{1\to2} - W_{3\to4}}{Q_{3\to4}} = 1 - \frac{T_Y}{T_X}. \tag{7.3}$$

7.2.3 Szilard Engine in Disguise

There is an information engine hidden in this heat engine. Imagine only having access to the X-position of each gas molecule during the cycle: In step $2 \to 3$ the container is divided and each gas molecule is either left or right of the partition. In step $3 \to 4$ input heat drives an isothermal expansion, and work is extracted from the gas as the piston is moved across the correct respective empty half of the container (appearing to act on hidden knowledge of the system state): It seems as if heat is entirely converted into work. However, with full access to the state space we recognize that this expansion step, seemingly requiring hidden knowledge, is preceded by a compression step that correlates each molecule's Y position with its X position, and the expansion step utilizes the Y position as a memory to execute the expansion into the correct half.

Imagining this process with a single molecule, we obtain the famous Szilard engine [51, 52], arguably the simplest information engine used to illustrate the Maxwell-demon paradox [53]. For the Szilard engine, the paradox is resolved by explicitly accounting for the memory degree of freedom and using Landauer's principle [24] to show that information erasure carries a thermodynamic cost. In analogy to this, we analyze the information thermodynamics [54] of the interplay between the X and Y components of the gas.

In the initial, uncompressed state the N-molecule gas has joint entropy

$$S_1[X, Y] = N \ln V + \frac{N}{2} \ln T_X + \frac{N}{2} \ln T_Y , \tag{7.4}$$

up to constants irrelevant to the analysis. (Note that while we appear to take logarithms of dimensional quantities, this is because we have omitted additional constant factors [55].) Using the container's side length L (and hence volume $V = L^2$), the joint entropy is

$$S_1[X, Y] = S_1[X] + S_1[Y] , \tag{7.5a}$$

for marginal entropies

$$S_1[X] = N \ln L + \frac{N}{2} \ln T_X \tag{7.5b}$$

$$S_1[Y] = N \ln L + \frac{N}{2} \ln T_Y . \tag{7.5c}$$

Initially, at equilibrium without any coupling, the subsystems X and Y are independent and hence they have zero mutual information: $I_1[X; Y] = S_1[X] + S_1[Y] - S_1[X, Y] = 0$.

The compression step $1 \to 2$ does not change the one-dimensional phase space available to each individual component. Consequently, the marginal entropies remain unchanged: $S_2[X] = S_1[X]$ and $S_2[Y] = S_1[Y]$. Nonetheless, the joint entropy has been reduced by $N \ln 2$, since the (joint) phase-space volume available to each molecule has been halved. Since the joint entropy has been reduced while holding the marginal entropies constant, mutual information has been introduced between the components:

$$S_2[X, Y] = S_2[X] + S_2[Y] - I_2[X; Y] \tag{7.6a}$$

$$= S_1[X, Y] - N \ln 2 , \tag{7.6b}$$

and hence

$$I_2[X;Y] = N \ln 2. \tag{7.7}$$

This can be thought of as using a binary memory variable Y to encode each molecule's coarse-grained X-position: whether it is left or right of the divider. Therefore each molecule's coarse-grained Y-position (*above* or *below* the divider) acts as a memory of each molecule's coarse-grained X-position. Reducing entropy incurs a thermodynamic cost, expressed by the work done (or equivalently the heat flow) in this step,

$$W_{1\to 2} = -Q_{1\to 2} = k_B T_Y I_2[X;Y]. \tag{7.8}$$

Hence the work $W_{1\to 2}$ produces mutual information $I_2[X;Y]$.

Swapping the mobilities of the partitions leaves the entropy unaffected: $S_3[X;Y] = S_2[X,Y]$. The expansion step $3 \to 4$ returns the engine back to the initial state and thus

$$S_4[X,Y] = S_1[X,Y] = S_1[X] + S_1[Y]. \tag{7.9}$$

Therefore, the expansion "uses up" the previously created mutual information $I_2[X;Y]$ to do work:

$$W_{3\to 4} = -k_B T_X I_2[X;Y]. \tag{7.10}$$

From this point of view, the engine cycle consists of a measurement step $1 \to 2$ creating information, and feedback step $3 \to 4$ exploiting that information. Thereby the Y-component of the gas acts as the controller and memory of an information engine whose working substance is the X-component of the gas.

7.3 Bipartite Stochastic Thermodynamics and Information-Flow Arbitrage Relation

We now mathematically formalize the above ideas to demonstrate that any bipartite heat engine must utilize information to achieve positive output power. Specifically, we derive an inequality that bounds the output power of a bipartite heat engine by its information flow.

With the subsystems X and Y in contact with heat baths at different temperatures T_X and T_Y, the subsystem second laws (Eqs. (2.47)) are:

$$\dot{\Sigma}_X = \mathrm{d}_t S[X] - \beta_X \dot{Q}_X - \dot{I}_X \geq 0, \tag{7.11a}$$

$$\dot{\Sigma}_Y = \mathrm{d}_t S[Y] - \beta_Y \dot{Q}_Y - \dot{I}_Y \geq 0. \tag{7.11b}$$

In addition to the second laws (7.11), bipartite systems also satisfy a first law describing energy balance:

$$\dot{W} + \dot{Q}_X + \dot{Q}_Y = \mathrm{d}_t E. \tag{7.12}$$

Here E is the internal energy of the system, and $\dot{W}$ is the rate of work into the system, which in general may include contributions from both nonconservative driving forces and changes in potential energy due to varying external control parameters.

7.3.1 Nonequilibrium Steady States

We now consider the special case of autonomous systems not subject to time-dependent external control. At nonequilibrium steady state (NESS), the first and second laws combine to yield two inequalities constraining sums of the external work rates, the transduced power (energy flow), and the information flow:

$$\dot{\Sigma}_X = \beta_X \dot{W}_X + \beta_X \dot{E}_Y + \dot{I}_Y \geq 0\,, \tag{7.13a}$$

$$\dot{\Sigma}_Y = \beta_Y \dot{W}_Y - \beta_Y \dot{E}_Y - \dot{I}_Y \geq 0. \tag{7.13b}$$

Multiplying Eq. (7.13) by k_B and the respective subsystem temperatures T_X and T_Y, summing the two equations, and rearranging yields an upper bound on the total output power in terms of the temperature difference and the information flow:

$$-\dot{W} = -\dot{W}_X - \dot{W}_Y \leq k_\mathrm{B}(T_X - T_Y)\dot{I}_Y. \tag{7.14}$$

We call this result the *information-flow arbitrage relation* (*IFAR*), and it holds significant implications for the functioning of bipartite heat engines. It makes precise the notion, described in Sect. 7.2.1 through analogy to economic arbitrage, that information flows are necessary; to achieve net output work ($-\dot{W} > 0$), there must be a flow of information between the two subsystems. Moreover, it must be the colder subsystem whose dynamics increase correlations (and thus the hotter subsystem whose dynamics reduce them): if $T_X > T_Y$, then a functional heat engine must have $\dot{I}_Y > 0$ (and thus $\dot{I}_X < 0$).

In addition to showing that bipartite heat engines require information flows, the IFAR also provides a necessary condition for information engines to achieve net output power $-\dot{W} > 0$: a positive information flow $\dot{I}_Y$ is not sufficient; a temperature difference is also required. Only then does the controller at T_Y "pay" less energy to create correlations than the controlled system at $T_X > T_Y$ extracts by consuming those correlations. This reflects a fundamental connection between information engines and bipartite heat engines: to achieve net output power, an information engine must leverage fluctuations of different strengths (e.g., a temperature difference, which is the driving force for a heat engine), while a bipartite heat engine must contain an information flow (the hallmark of an information engine).

While the IFAR bounds the net output power $-\dot{W}$, with dimensions of energy divided by time, this output power is spread across two different temperatures. These different temperatures constitute different conversion rates between energy and entropy, and so the two output powers $-\dot{W}_X$ and $-\dot{W}_Y$ are each measured relative to different thermal backgrounds. We can take these different "exchange rates" into account by considering the sum of the two output powers scaled by their respective temperatures, $-\beta_X \dot{W}_X - \beta_Y \dot{W}_Y$. Similar to the derivation of the IFAR, summing Eq. (7.13) and rearranging yields

$$-\beta_X \dot{W}_X - \beta_Y \dot{W}_Y \leq (\beta_X - \beta_Y)\dot{E}_Y, \tag{7.15}$$

which we call the *transduced-power arbitrage relation* (*TPAR*). The TPAR states that to obtain a positive sum of scaled output powers, the hot (X) subsystem must transduce work to the cold (Y) subsystem so that $\dot{E}_Y < 0$.

An intriguing possible configuration of a bipartite heat engine is one where net output work is extracted from both subsystems (both $\dot{W}_X < 0$ and $\dot{W}_Y < 0$), as illustrated in Fig. 7.2c. (We will show an explicit example of such a heat engine in Sect. 7.4.1.) Combining the IFAR (7.14) and TPAR (7.15) shows that extracting net output work from both subsystems is only possible for specific directions of the internal energy and information flows. In particular, transduced work must flow from hot to cold ($\dot{E}_Y < 0$) while information must flow from cold to hot ($\dot{I}_Y > 0$).

7.3.2 Interpretation in Terms of Environmental Potentials

For an alternative interpretation of these results, consider the thermodynamics from the perspective of the environment, comprised of two reservoirs at T_X and T_Y respectively. These reservoirs are assumed to be at equilibrium, so each have well-defined energies ($\mathcal{U}_X$ and $\mathcal{U}_Y$) and entropies ($\mathcal{S}_X$ and $\mathcal{S}_Y$). From these state functions we can construct thermodynamic potentials for the environment, for example the Helmholtz free energy

$$\mathcal{F} \equiv \mathcal{U}_X - k_B T_X \mathcal{S}_X + \mathcal{U}_Y - k_B T_Y \mathcal{S}_Y, \tag{7.16}$$

and free entropy (also known as the Massieu potential [55])

$$\Phi \equiv \mathcal{S}_X - \beta_X \mathcal{U}_X + \mathcal{S}_Y - \beta_Y \mathcal{U}_Y. \tag{7.17}$$

Since the reservoirs interact (and thus exchange energy and entropy) only with their respectively coupled subsystems, the rates of change of their energies and entropies are $\dot{\mathcal{U}}_X = -\dot{W}_X - \dot{Q}_X$ and $\dot{\mathcal{S}}_X = -\beta_X \dot{Q}_X$, and likewise for $\dot{\mathcal{U}}_Y$ and $\dot{\mathcal{S}}_Y$. We can then compute the steady-state rates of change of the two environmental potentials:

$$\dot{\mathcal{F}} = -\dot{W}_X - \dot{W}_Y, \tag{7.18a}$$

$$\dot{\Phi} = \beta_X \dot{W}_X + \beta_Y \dot{W}_Y. \tag{7.18b}$$

We then substitute these definitions into the left hand sides of the IFAR (7.14) and TPAR (7.15) to reformulate them in terms of rates of change of environmental potentials:

$$\dot{\mathcal{F}} \leq k_B(T_X - T_Y)\dot{I}_Y, \tag{7.19a}$$

$$-\dot{\Phi} \leq (\beta_X - \beta_Y)\dot{E}_Y. \tag{7.19b}$$

These reformulations lead to a new interpretation with a pleasing symmetry. The IFAR states that the rate at which the system can leverage a temperature difference to increase the free energy of the environment is limited by the rate of internal information (entropy) transduction, while conversely the TPAR states that the rate at which the system can decrease the free entropy of the environment is limited by the rate of internal energy transduction.

Intuitively, these results follow from the definitions of environmental free energy (7.16) and free entropy (7.17). Since the energetic terms of the free energy (7.16) are not modulated by temperature, and at steady state their sum must remain constant, it follows that $\mathcal{F}$ can only be increased by moving entropy from the hotter reservoir to the colder one. The reservoirs only interact indirectly via the system, which must then serve as a conduit for the entropy flow, which we call an information flow. Thus environmental free energy can only increase if the system supports an internal information flow.

Conversely, the entropic terms of the free entropy (7.17) are not modulated by temperature, and at steady state cannot decrease (by the second law), so it follows that Φ can only be decreased by moving energy from the hotter reservoir to the colder one. As with entropy, energy can only be exchanged using the system as a conduit. Thus decreasing the environmental free entropy requires an internal energy flow, i.e., a transduced power from hot to cold.

This interpretation of the two arbitrage relations is intimately connected with the economics analogy outlined earlier in Sect. 7.2. A bipartite heat engine which increases the free energy of the environment corresponds to a pair of arbitrageurs who cooperate to extract net money from two markets with different exchange rates; as illustrated in Fig. 7.2 this is only possible when the two arbitrageurs exchange sheep (i.e., when the heat engine supports an information flow). Conversely, decreasing the free entropy of the environment corresponds to the arbitrageurs obtaining a net return of sheep, which in turn requires them to exchange money with each other (corresponding to the two subsystems supporting a transduced power).

7.3.3 Connection to Carnot Bound

The IFAR (7.14) relates the output work of a bipartite heat engine to the information flow and temperature difference. The output work of a heat engine operating between two reservoirs was famously first upper-bounded by Carnot [56], in terms of the temperature ratio and the input heat. Because in a bipartite heat engine, each of the two reservoirs is coupled to a distinct individual subsystem, the input heat from the "hot" reservoir can only flow into the "hot" subsystem X. Then, the Carnot bound for heat engines at steady state is derived from the global first law,

$$\dot{W} + \dot{Q}_X + \dot{Q}_Y = \mathrm{d}_t E \tag{7.20}$$

(with $\mathrm{d}_t E = 0$), and the global second law,

$$\dot{\Sigma} = -\beta_X \dot{Q}_X - \beta_Y \dot{Q}_Y \geq 0. \tag{7.21}$$

This is simply a sum of the two subsystem-specific second laws (7.11). Rearranging Eq. (7.21) to get an upper bound on $\dot{Q}_Y$ and inserting into Eq. (7.20) yields the Carnot bound on the ouput power of a heat engine:

$$-\dot{W} \leq \left(1 - \frac{T_Y}{T_X}\right) \dot{Q}_X. \tag{7.22}$$

The Carnot bound essentially states that in a heat engine input heat limits output work, with a proportionality constant dependent on the temperature ratio.

Now notice that we have $\dot{Q}_X \geq 0$, i.e., heat flows into subsystem X (which, recall, is coupled to only one reservoir), something seemingly forbidden by the second law and the hallmark of a Maxwell demon [57–60]. Under steady-state conditions, the second law for the X subsystem (7.11a) can be rewritten to bound the achievable input heat by the information flow,

$$\dot{Q}_X \leq k_\mathrm{B} T_X \dot{I}_Y. \tag{7.23}$$

Inserting this inequality into the Carnot bound yields the information-flow arbitrage relation:

$$\underbrace{\overbrace{-\dot{W} \leq \left(1 - \frac{T_Y}{T_X}\right) \dot{Q}_X}^{\text{Carnot}} \underbrace{\leq k_\mathrm{B}(T_X - T_Y)\dot{I}_Y}_{\text{2nd Law for } X}}_{\text{IFAR}}. \tag{7.24}$$

Thus we find that the IFAR is in general looser than the Carnot bound on heat-engine output work. Nonetheless, IFAR broadens the perspective by showing that working bipartite heat engines necessarily require an information flow. Note also that IFAR is saturated for heat engines at equilibrium (where output work vanishes), as is the Carnot bound [61].

7.3.4 Periodic Driving

While biological systems of interest operate autonomously and are typically in nonequilibrium steady states, many human-engineered systems (e.g., classical heat engines and the 2D ideal gas considered in Sect. 7.2.2) are controlled by periodic driving protocols. Likewise, experimental [62–73] and theoretical models [42, 74–78] of information engines typically use repeated feedback loops [25, 79–84] which comprise measurement, feedback, and relaxation steps. These setups can also be understood in terms of a periodic driving protocol that achieves the desired feedback [23]. Information engines with access to a temperature difference [41, 42] or nonequilibrium fluctuations that only affect their working medium and not their controller [36–38] can have positive net work output. Here we show that the IFAR (7.14) also holds for this class of systems.

For bipartite systems in periodic steady states, the IFAR is derived analogously to systems at NESS. We take as a starting point the instantaneous second laws (7.11), and integrate over the cycle time τ to yield their cyclic counterparts:

$$\Sigma_X = \Delta S[X] - \beta_X Q_X - \Delta I_X \geq 0, \tag{7.25a}$$

$$\Sigma_Y = \Delta S[Y] - \beta_Y Q_Y - \Delta I_Y \geq 0. \tag{7.25b}$$

Here $Q_X \equiv \int_0^\tau \dot{Q}_X(t)\,dt$, $\Sigma_X \equiv \int_0^\tau \dot{\Sigma}_X(t)dt$, $\Delta S[X] \equiv S_\tau[X] - S_0[X]$, and $\Delta I_X \equiv \int_0^\tau \dot{I}_X dt$, with analogous definitions holding for flows due to Y. Likewise, the global first law also integrates to yield

$$W + Q_X + Q_Y = \Delta E, \tag{7.26}$$

where $W \equiv \int_0^\tau \dot{W} dt$ and $\Delta E \equiv E(\tau) - E(0)$.

For systems at periodic steady states, we require that all state variables are identical at the beginning and end of the cycle. Specifically these are the mean internal energy $[\Delta E = E(\tau) - E(0) = 0]$, the two marginal entropies ($\Delta S[X] = S_\tau[X] - S_0[X] = 0$ and $\Delta S[Y] = S_\tau[Y] - S_0[Y] = 0$), and the mutual information ($\Delta I_X + \Delta I_Y = I_\tau[X;Y] - I_0[X;Y] = 0$). Using these invariants to simplify the first (7.26) and second laws (7.25), we derive the IFAR for periodically driven systems:

$$-W \leq k_B(T_X - T_Y)\Delta I_Y. \tag{7.27}$$

The TPAR (7.15) does not extend as easily to periodic steady states, since in the presence of external control, defining external work at the subsystem level requires a more nuanced analysis beyond the scope of this thesis.

7.3.5 Revisiting the 2D Ideal-Gas Engine

Having derived the IFAR for periodically driven systems, we return to the 2D ideal-gas engine in Sect. 7.2.2 to illustrate this relation.

Intuitively, the total Y-information flow over one cycle should be given in terms of the information acquired during the compression ($1 \to 2$) step, when the molecules' Y-coordinates change. To see why this is indeed true, consider first the *coarse-grained* variables $\bar{X}^i := \text{sgn}(X^i)$ and $\bar{Y}^i := \text{sgn}(Y^i)$ that respectively indicate whether a given molecule i is left or right of and above or below the box's center. The gas molecules' coordinates are in equilibrium hence the conditional probability of the specific position given the coarse-grained position is uniform and factorizes,

$$p(X^i, Y^i|\bar{X}^i, \bar{Y}^i) = p(X^i|\bar{X}^i)\, p(Y^i|\bar{Y}^i). \tag{7.28}$$

This implies for the conditional entropy of the specific position given the coarse-grained position, $S[X^i, Y^i|\bar{X}^i, \bar{Y}^i] = S[X^i|\bar{X}^i] + S[Y^i|\bar{Y}^i]$. Therefore, the mutual information between the coordinates X^i and Y^i equals the mutual information between the coarse-grained coordinates $\bar{X}^i$ and $\bar{Y}^i$,

$$
\begin{aligned}
I[X^i; Y^i] &= S[X^i] + S[Y^i] - S[X^i, Y^i] && (7.29a)\\
&= S[X_i, \bar{X}_i] + S[Y^i, \bar{Y}^i] - S[X^i, Y^i, \bar{X}^i, \bar{Y}^i] && (7.29b)\\
&= S[\bar{X}^i] + S[\bar{Y}^i] - S[\bar{X}^i, \bar{Y}^i] + \underbrace{S[X^i|\bar{X}^i] + S[Y^i|\bar{Y}^i] - S[X^i, Y^i|\bar{X}^i, \bar{Y}^i]}_{=0} && (7.29c)\\
&= I[\bar{X}^i; \bar{Y}^i], && (7.29d)
\end{aligned}
$$

where to get (7.29b) we used the fact that once a molecule's true position (X^i, Y^i) is known, the coarse-grained position $(\bar{X}^i, \bar{Y}^i)$ is redundant information. From Eqs. (7.29), it follows that to calculate information flow, we can replace all mutual information terms by their coarse-grained counterparts.

The total Y-information flow is

$$
\begin{aligned}
\Delta I_Y &= \int_1^2 \mathrm{d}t\, \dot{I}_Y + \int_3^4 \mathrm{d}t\, \dot{I}_Y && (7.30a)\\
&= \int_1^2 \mathrm{d}t \lim_{\mathrm{d}t\to 0} \frac{I[\bar{X}(t+\mathrm{d}t); \bar{Y}(t+\mathrm{d}t)] - I[\bar{X}(t); \bar{Y}(t)]}{\mathrm{d}t}\\
&\quad + \int_3^4 \mathrm{d}t \lim_{\mathrm{d}t\to 0} \frac{I[\bar{X}(t); \bar{Y}(t)] - I[\bar{X}(t); \bar{Y}(t)]}{\mathrm{d}t} && (7.30b)\\
&= I_2[\bar{X}; \bar{Y}] - \underbrace{I_1[\bar{X}; \bar{Y}]}_{=0} = I_2[X; Y], && (7.30c)
\end{aligned}
$$

where $I_2[X; Y]$ is the mutual information between the molecules' X and Y coordinates after compression, see Eq. (7.7). To get Eq. (7.30b) we used the definition of the Y-information flow (2.51) and the fact that $\bar{X}^i(t + \mathrm{d}t) = \bar{X}^i(t)$ in step 1→2 and $\bar{Y}^i(t + \mathrm{d}t) = \bar{Y}^i(t)$ in step 3→4, i.e., during compression (expansion) the coarse-grained $X(Y)$-coordinate does not change. Similarly,

$$\Delta I_X = I_4[\bar{X}; \bar{Y}] - I_3[\bar{X}; \bar{Y}] = -I_2[X; Y] = -\Delta I_Y. \tag{7.31}$$

With the total net output work in Eq. (7.2) and the mutual information $I_2[X; Y] = N \ln 2$ (7.7), we verify that the IFAR (7.27) holds as an equality, which can be attributed to the fact that all steps in the cycle are carried out reversibly.

7.4 Model Systems: Brownian Gyrator and Double Quantum Dot

Using the theory of bipartite stochastic thermodynamics outlined above, we now analyze explicit nonequilibrium models to illustrate the main results: we consider the Brownian-gyrator heat engine [85] modified to incorporate nonconservative driving forces, and a double quantum dot which constitutes a simple model for an autonomous information engine. These models are analytically tractable, allowing us to show explicitly that the Brownian gyrator must use information flow to achieve net output power, and the quantum-dot information engine must act as a heat engine to deliver positive output power.

7.4.1 Brownian-Gyrator Heat Engine

Consider the Brownian gyrator, a microscopic, stochastic model for a steady-state heat engine depicted in Fig. 7.4a. First introduced in a slightly different form by Filliger and Reimann [85], the Brownian gyrator has since been studied extensively both in its original formulation [26, 86–89] along with a plethora of different extensions including the addition of conservative [90, 91] or nonconservative [92, 93] external forces, higher-order potentials [94], underdamped dynamics [95], as well as non-Markovian [96] and active [97] fluctuations. The dynamics of the Brownian gyrator can also be mapped directly onto electric-circuit models with resistors subject to Johnson noise from different heat baths; this equivalent system and several extensions have been thoroughly studied both theoretically and experimentally [98–102]. The Brownian gyrator has also been realized experimentally as an overdamped particle with electromagentically induced anisotropic fluctuations [103, 104].

This formulation of the Brownian gyrator, illustrated in Fig. 7.4a, is equivalent to that presented in Ref. [92], a bipartite system with two degrees of freedom X and Y whose dynamics evolve according to the coupled overdamped Langevin equations

$$\dot{x} = ay - \partial_x V(x, y) + \sqrt{2}\,\xi_X(t), \tag{7.32a}$$

$$\dot{y} = g\left[-ax - \partial_y V(x, y)\right] + \sqrt{2g\tau}\,\xi_Y(t). \tag{7.32b}$$

Here $\xi_X(t)$ and $\xi_Y(t)$ are uncorrelated Gaussian white noise sources with $\langle \xi_X(t)\xi_X(t')\rangle = \delta(t - t')$ and similarly for $\xi_Y(t)$, g is the ratio of the two mobility coefficients, $\tau \equiv T_Y/T_X$ is the temperature ratio, and the potential is

$$V(x, y) = \frac{1}{2}x^2 + \frac{1}{2}y^2 + \frac{1}{2}\kappa(x - y)^2, \tag{7.33}$$

for coupling strength κ between X and Y. Furthermore, $f_X(x, y) = ay$ and $f_Y(x, y) = -ax$ are nonconservative forces of strength a that, on their own, induce a rotation of the system in the $x-y$ plane. All quantities are dimensionless. The nonconservative forces were not present in the original formulation of the Brownian gyrator [85], and have been added here as in Ref. [92] so that work can be input to or extracted from the Brownian gyrator in a thermodynamically consistent manner. Such nonconservative forces can be incorporated into electrical implementations of the gyrator through, for example, a non-reciprocal capacitor [102].

The coupled Langevin equations (7.32) are linear, so the stationary joint probability distribution for X and Y can be solved analytically [105], from which the ensemble-averaged energy and information flows can be computed analytically from their definitions [35] using Maple [106] (plotted in Fig. 7.4).

Figure 7.4b illustrates the energy and information flows in the Brownian gyrator. In the regime of heat-engine operation, when $0 < a < \kappa(1 - \tau)/(1 + \tau)$, both subsystems ($X$ and Y) output work at positive rates ($\dot{W}_X < 0$ and $\dot{W}_Y < 0$). This is powered by a flow of heat into the hotter X subsystem ($\dot{Q}_X > 0$), which by the second law then requires a net flow of heat out of the cooler Y subsystem ($\dot{Q}_Y < 0$). As required by the arbitrage relations (7.14) and (7.15), the information flow and transduced power are both non-zero, with information flowing from cold to hot ($\dot{I}_X < 0$) and transduced work flowing from hot to cold ($\dot{E}_X > 0$). This thermodynamic setup is shown in Fig. 7.4c.

For $\kappa(1 - \tau)/(1 + \tau) < a < \kappa$, the Brownian gyrator operates as a heat pump, with net input work into both subsystems ($\dot{W}_X > 0$ and $\dot{W}_Y > 0$) powering the flow of heat from cold to hot ($\dot{Q}_Y > 0$ and $\dot{Q}_X < 0$). Here the information flow goes from hot to cold ($\dot{I}_X > 0$), a condition which follows directly from the second law applied to bipartite heat pumps. For $-\kappa < a < 0$ the Brownian gyrator is a dud, with input work into both subsystems accompanying a flow of heat from hot to cold. Finally for $|a| > \kappa$, the setup is again a dud (albeit of a different sort), with input work into both subsystems resulting in heat flows into both reservoirs.

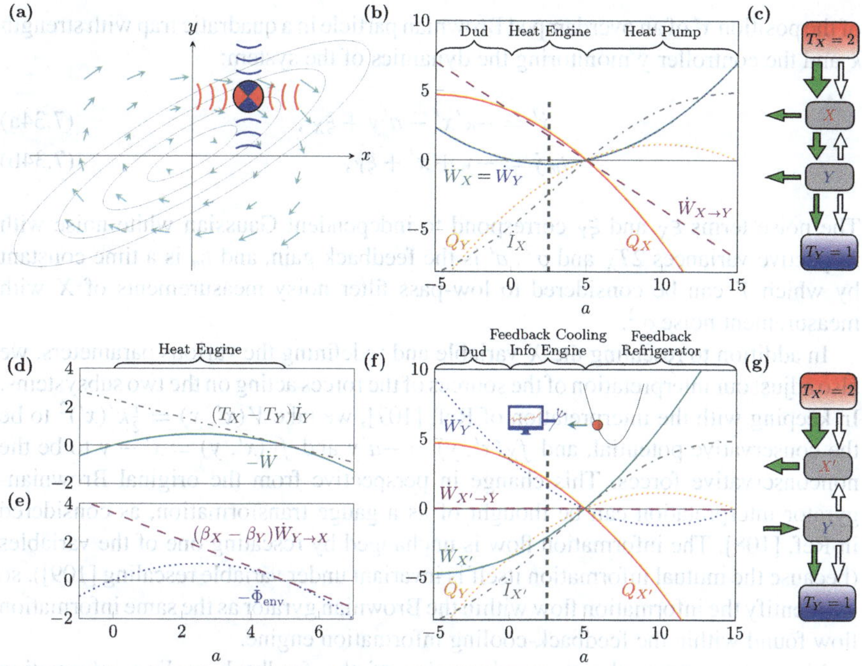

Fig. 7.4 Dynamics and thermodynamics of the Brownian gyrator. **a** Schematic diagram. Gray closed curves denote isopotential contours; light green arrows show the nonconservative force field; red-blue circle denotes the 2D Brownian particle; blue and red arcs denote different-strength fluctuations in different directions. **b** Energy and information flows. **c** Thermodynamic diagram showing directionality of energy, entropy, and information flows for $a = 2.5$ (black dashed vertical line in (**b**)). **d**, **e** Verification of the two arbitrage relations, IFAR (7.14) (d) and TPAR (7.15) (e). **f** Energy and information flows in the feedback-cooling information-engine interpretation. Inset: schematic, with nonconservative force from controller indicated by green arrow. **g** Thermodynamic diagram showing directionality of energy, entropy, and information flows for $a = 2.5$ (black dashed vertical line in (**f**)). Throughout, $\kappa = 15$, $\tau = 1/2$, and $g = 1$. Copyrighted by the American Physical Society [43]

Figure 7.4d and e explicitly confirm the two arbitrage relations in this system. We find that the inequalities are tighter when the system is near equilibrium. For the Brownian gyrator this is at stall, where all energy and information flows vanish, for $a = \kappa(1 - \tau)/(1 + \tau)$.

If, as we claim, information flow along with a temperature difference is what drives the net power output in the Brownian gyrator, we should be able to find the information engine hidden in this setup. To do this, we rescale the variable X such that $x' \equiv x(\kappa - a)/(1 + \kappa)$, and define the parameters $\kappa' \equiv 1 + \kappa$, $t_m \equiv 1/[g(1 + \kappa)]$, $\sigma^2 \equiv 2\tau/[g(1 + \kappa)^2]$, and $a' \equiv (a^2 - \kappa^2)/(1 + \kappa)$.

The resulting dynamics are equivalent to the feedback-cooling information-engine model studied by Horowitz and Sandberg [107], with the controlled system consisting of the position x' of an overdamped Brownian particle in a quadratic trap with strength κ and the controller y monitoring the dynamics of the system:

$$\dot{x}' = -\kappa' x' - a' y + \xi_{X'}, \tag{7.34a}$$

$$t_m \dot{y} = -y + x' + \xi_Y, \tag{7.34b}$$

The noise terms ξ_X and ξ_Y correspond to independent Gaussian white noise with respective variances $2T_X$ and σ^2, a' is the feedback gain, and t_m is a time constant by which Y can be considered to low-pass filter noisy measurements of X with measurement noise σ^2.

In addition to rescaling the X variable and redefining the various parameters, we also adjust our interpretation of the sources of the forces acting on the two subsystems. In keeping with the interpretation of Ref. [107], we take $V(x', y) = \frac{1}{2}\kappa'(x')^2$ to be the conservative potential, and $f_X(x', y) = -a'y$ and $f_Y(x', y) = x' - y$ to be the nonconservative forces. This change in perspective from the original Brownian-gyrator interpretation can be thought of as a gauge transformation, as considered in Ref. [108]. The information flow is unchanged by rescaling one of the variables (because the mutual information itself is invariant under variable rescaling [109]), so we identify the information flow within the Brownian gyrator as the same information flow found within the feedback-cooling information engine.

Figure 7.4f shows the thermodynamics of the feedback-cooling information engine, as quantified by the energy and information flows. The heat and information flows are unchanged from those of the Brownian gyrator, but the input, output, and transduced powers are modified. In particular, since there is no potential energy coupling the controller to the particle, the transduced power $\dot{E}_X$ is always zero. For $a' < \kappa'(1 - \tau)/(1 + \tau)$, output work is extracted from the particle ($\dot{W}_X < 0$), while input work is required to run the controller ($\dot{W}_Y > 0$). The three thermodynamic regimes highlighted for the Brownian gyrator in Fig. 7.4b correspond to matching regimes for the information engine depicted in Fig. 7.4f. In particular, the heat-engine regime corresponds to a feedback-cooling information engine that extracts more power than is required to power it, while the dud regime corresponds to an information engine that consumes more energy than it extracts. Finally, the heat-pump regime corresponds to a feedback refrigerator, where the controller Y is cooled through feedback on the hotter system X'.

7.4.2 *Double-Quantum-Dot Information Engine*

The above treatment of the Brownian gyrator focused on interpreting a bipartite heat engine as an information engine. This section illustrates the converse, showing that an established model of an autonomous information engine can only deliver positive total output power if the controller and feedback-controlled system are at different temperatures, thus rendering the system a bipartite heat engine.

Consider a single quantum dot X in contact with two reservoirs (leads) at temperature T_X and with chemical potentials μ_ℓ for the left lead and μ_r for the right lead [Fig. 7.5a inset]. Electrons can jump between either reservoir and the quantum dot, which can be either empty ($x = 0$) or filled ($x = 1$). The rates satisfy detailed balance with their respective thermal reservoirs:

$$k_\ell^{10} = \frac{\Gamma_X^\ell}{e^{-\beta_X \mu_\ell} + 1}, \quad k_\ell^{01} = \frac{\Gamma_X^\ell \, e^{-\beta_X \mu_\ell}}{e^{-\beta_X \mu_\ell} + 1} \tag{7.35a}$$

$$k_\mathrm{r}^{10} = \frac{\Gamma_X^\mathrm{r}}{e^{-\beta_X \mu_\mathrm{r}} + 1}, \quad k_\mathrm{r}^{01} = \frac{\Gamma_X^\mathrm{r} \, e^{-\beta_X \mu_\mathrm{r}}}{e^{-\beta_X \mu_\mathrm{r}} + 1} \tag{7.35b}$$

where $\Gamma_X^\ell/2$ and $\Gamma_X^\mathrm{r}/2$ are bare rate constants, i.e., the rate constants at equilibrium (when $\mu_\ell = 0 = \mu_\mathrm{r}$). Here and throughout, the superscript "10" denotes the transition from state 0 to state 1, with "01" denoting the reverse transition. Note that in this subsection the symbol W with a subscript and a superscript denotes a transition rate, not to be confused with $\dot{W}$ with a subscript that throughout this thesis denotes power.

The average current of electrons flowing from the left lead is $J_\ell \equiv k_\ell^{10} p_0 - k_\ell^{01} p_1$, where p_0 and p_1 are the respective stationary probabilities for the dot being empty and filled. Similarly, $J_\mathrm{r} \equiv k_\mathrm{r}^{01} p_1 - k_\mathrm{r}^{10} p_0$. Solving the system of equations consisting of the definitions of J_ℓ and J_r, the steady-state equality $J_\ell = J_\mathrm{r} = J$, and normalization $p_0 + p_1 = 1$, yields p_0, p_1, and J as functions of the chemical potentials μ_ℓ and μ_r, temperature T_X, and bare rate constants Γ_X^r and Γ_X^ℓ. These then allow us to calculate the steady-state power from the quantum dot to the reservoirs,

$$-\dot{W}_X \equiv J\,(\mu_\mathrm{r} - \mu_\ell)\,. \tag{7.36}$$

This power is equal to the heat flow from the thermal reservoir due to the conduction of electrons through the quantum dot. Figure 7.5a depicts this power as a function of the chemical potential μ_r for fixed μ_ℓ, T_X, and $\Gamma_X^{\ell,\mathrm{r}}$.

"Maxwell-demon feedback" [110] can be used to pump electrons from right to left, against the chemical-potential gradient. Like Maxwell's original demon [111] that implements feedback by opening and closing a trap door between two gas volumes, this feedback only modifies the bare transition rates, not the energy levels. We assume that the state of the quantum dot is continuously measured by an auxiliary measurement device that modifies the bare rates Γ_X such that $\Gamma_X^\ell = 0$ when the dot is measured empty and $\Gamma_X^\mathrm{r} = 0$ when the dot is measured filled [Fig. 7.5b inset].

The measurement has an error probability ϵ with which it erroneously measures the opposite state. Then the net flux from left to right is

$$J = k_\ell^{10} p_0 \epsilon - k_\ell^{01} p_1 (1 - \epsilon) \tag{7.37a}$$

$$= k_\mathrm{r}^{01} p_1 \epsilon - k_\mathrm{r}^{10} p_0 (1 - \epsilon). \tag{7.37b}$$

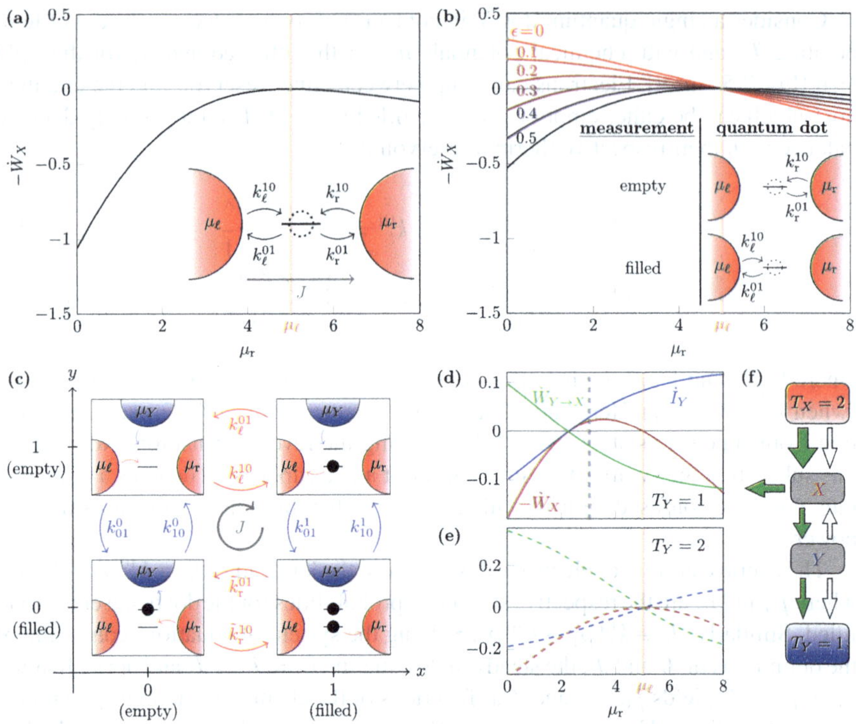

Fig. 7.5 Thermodynamics of the quantum-dot information engine. **a** Output power $-\dot{W}_X$ by a single quantum dot in simultaneous contact with two reservoirs at temperature $T_X = 2$ and respective chemical potentials $\mu_\ell = 5$ and μ_r, as a function of μ_r, for $\Gamma_X^\ell = \Gamma_X^r = 1$. Inset: transition rates for electrons hopping into and out of the quantum dot. **b** When measurement with error ϵ and feedback are added, $-\dot{W}_X > 0$ is possible, indicating that electrons are pumped against their natural gradient. **c** State diagram and possible transitions for the double-quantum-dot model. **d** Output power $-\dot{W}_X$, transduced power $\dot{E}_Y$, and information flow $\dot{I}_Y$ in the double quantum dot for the same parameters as in (**a**) and (**b**), and $\epsilon = 0.2$, $T_Y = 1$, $\Gamma_Y = 10^3$. **e** Same plot but for $T_Y = 2$. **f** Thermodynamic diagram showing directionalities of energy, entropy, and information flows for $\mu_r = 3$ and $T_Y = 1$ [gray dashed vertical line in (**d**)]. Throughout: faint vertical orange line indicates μ_ℓ. Copyrighted by the American Physical Society [43]

Solving this equation with normalization $p_0 + p_1 = 1$ yields the steady-state occupation probabilities and the flux as functions of μ_ℓ, μ_r, T_X, and ϵ. Equation (7.36) then gives the steady-state output power $-\dot{W}_X$ [Fig. 7.5b]. For $\epsilon < 1/2$, this power can become positive, indicating that the quantum dot delivers net power to the reservoirs. This output power stems from the heat due to thermal fluctuations that spontaneously fill the dot with an electron from the right reservoir. The feedback then rectifies these fluctuations by preferentially allowing the electron to flow into the left reservoir.

The quantum dot and the feedback mechanism together constitute a Maxwell-demon setup or information engine. However, we have yet to specify the physical mechanism of the controller. As pointed out in previous works [18, 57, 112–115], a second capacitively coupled quantum dot can serve as the controller for this information engine. Consider the single-level quantum dot Y that is coupled to a reservoir with chemical potential μ_Y and temperature T_Y. Somewhat counterintuitively, a filled Y ($y = 0$) can encode the measurement of an empty X ($x = 0$) and an empty Y ($y = 1$) can encode the measurement of a filled X ($x = 1$). The capacitive interaction between the dots results in a potential energy U when both dots are filled. Otherwise, the potential energy vanishes.

Figure 7.5c depicts the transition rates between the four different configurations of the double quantum dot. The transition rates into and out of the X-dot are

$$k_y^{xx'} = \begin{cases} \tilde{k}_{\mathrm{r}}^{xx'}, & y = 0 \text{ (filled)} \\ k_\ell^{xx'}, & y = 1 \text{ (empty)}, \end{cases} \tag{7.38}$$

where x and x' are the states of X with $x \neq x'$, $k_\ell^{xx'}$ is given by Eq. (7.35a), and

$$\tilde{k}_{\mathrm{r}}^{10} = \frac{\Gamma_X^{\mathrm{r}}}{e^{-\beta_X(\mu_{\mathrm{r}}-U)} + 1}, \tag{7.39a}$$

$$\tilde{k}_{\mathrm{r}}^{01} = \frac{\Gamma_X^{\mathrm{r}}\, e^{-\beta_X(\mu_{\mathrm{r}}-U)}}{e^{-\beta_X(\mu_{\mathrm{r}}-U)} + 1} \tag{7.39b}$$

correspond to the rates in Eq. (7.35b), modified by the potential energy U due to the interaction with the other quantum dot Y. Equation (7.38) implies that the X-dot couples to different reservoirs depending on the state of the Y-dot. This is a special case of the treatment in, e.g., Refs. [18, 57] in which the bare rates Γ_X^ℓ and Γ_X^{r} are modified through a Y-dependent density of states. This model thus corresponds to an idealized double-quantum-dot information engine with only one global cycle and no local cycles [18].

To implement the measurement, the Y-transitions are governed by

$$k_{01}^{0} = \frac{\Gamma_Y}{e^{-\beta_Y \mu_Y} + 1}, \qquad k_{10}^{0} = \frac{\Gamma_Y\, e^{-\beta_Y \mu_Y}}{e^{-\beta_Y \mu_Y} + 1} \tag{7.40a}$$

$$k_{01}^{1} = \frac{\Gamma_Y}{e^{-\beta_Y(\mu_Y-U)} + 1}, \quad k_{10}^{1} = \frac{\Gamma_Y\, e^{-\beta_Y(\mu_Y-U)}}{e^{-\beta_Y(\mu_Y-U)} + 1} \tag{7.40b}$$

where $U = 2\mu_Y$ and $\mu_Y = k_{\mathrm{B}} T_Y \ln[(1-\epsilon)/\epsilon]$ are chosen such that

$$\frac{k_{01}^{0}}{k_{10}^{0}} = \frac{k_{10}^{1}}{k_{01}^{1}} = \frac{1-\epsilon}{\epsilon}, \tag{7.41}$$

and hence the measurement-error probability is ϵ. Setting $\Gamma_Y \gg \Gamma_X^{\mathrm{r},\ell}$ ensures that Y quickly relaxes into a local equilibrium distribution corresponding to the desired measurement distribution.

The net current of electrons from the left to right leads is determined by solving

$$
\begin{aligned}
J &= k_\ell^{10} p_{01} - k_\ell^{01} p_{11} = W_{01}^{1} p_{11} - k_{10}^{1} p_{10} \\
&= \tilde{k}_{\mathrm{r}}^{01} p_{10} - \tilde{k}_{\mathrm{r}}^{10} p_{00} = W_{10}^{0} p_{00} - k_{01}^{0} p_{01}
\end{aligned}
\tag{7.42a}
$$

along with normalization $p_{00} + p_{10} + p_{10} + p_{11} = 1$ constraining the probability p_{xy} to find the joint system in state (x, y). Figure 7.5d shows the net power (7.36) done on X by the two X-reservoirs.

Importantly, the output power $-\dot{W}_X$ differs from the case where the controller does not require power [see, for example, the $\epsilon = 0.2$ curve in Fig. 7.5b]. Specifically, positive output power ($-\dot{W}_X > 0$) is only possible for $2 \lesssim \mu_{\mathrm{r}} \lesssim 5$. This is because power is required to run the controller Y, which itself is not directly driven by a chemical-potential difference since it has access to only one reservoir, and thus $\dot{W}_Y = J\,(\mu_Y - \mu_Y) = 0$. As depicted in Fig. 7.5f, the power to run the controller Y is diverted from the output power as transduced power $-\dot{E}_Y$ [Fig. 7.5d, green curve]. The blue curve in Fig. 7.5d shows the information flow $\dot{I}_Y$, with which we can verify the IFAR (7.14),

$$
-\dot{W}_X - \underbrace{\dot{W}_Y}_{=0} \le \underbrace{(T_X - T_Y)}_{=1}\,\dot{I}_Y. \tag{7.43}
$$

Finally, Fig. 7.5e shows output power $-\dot{W}_X$ and transduced power $\dot{E}_Y$ when both quantum dots are at equal temperature ($T_Y = T_X = 2$), so the RHS of (7.43) vanishes and $-\dot{W}_X \le 0$. In this case so much power $-\dot{E}_Y$ is diverted that no positive output power can be generated. This illustrates that the double-quantum-dot information engine can only deliver positive output power when the controller is at a lower temperature than the feedback-controlled system, i.e., when the joint system operates as a bipartite heat engine, as predicted by the information-flow arbitrage relation (7.14).

7.5 Discussion

This chapter illustrates that functioning bipartite heat engines must transmit information (i.e., a reduction in entropy) between subsystems in contact with heat reservoirs at different temperatures in order to produce net output work, through a process analogous to economic arbitrage. This implies that they are also information engines, in the sense that they sustain an information flow that powers an apparent violation of the second law. This implies that the field of information thermodynamics [54] applies to real-world heat engines.

These findings directly imply design principles for nanoscale systems, like molecular machines, operating in environments with inhomogeneous or anisotropic fluctuations. When these systems are composed of different parts each in contact with different strengths of fluctuations, maximizing output power requires these components to operate collectively and exchange entropy in the form of information flows. This leads to "Maxwell-demon" behaviour, where one component extracts heat from its environment in apparent, but not true, violation of the second law. Thus Maxwell's demon may well lie hidden within biological molecular machines which have evolved to take advantage of different sources of fluctuations in the cellular environment.

Conversely, this chapter also illustrates that information engines can deliver positive net output power when controller and controlled system are at different temperatures. This fact was explored in Refs. [41, 42], where the ratio of the temperatures of the work medium and memory parameterizes optimal information-processing strategies in variants of the Szilard engine. It is also implicit in the analysis of Ref. [38], where an information engine delivers net power derived from active fluctuations that mimic the effect of a larger temperature.

> **Fundamental Limit:** The net output work of a bipartite heat engine or information engine is limited by the product of the information flow and temperature difference.
> **Design Principle:** A bipartite heat engine with maximal output power should be designed to function like an information engine.

This framework helps to demystify information engines by providing a change in perspective that illustrates they are variants of heat engines, in which the entropy reduction step is "outsourced" to an auxiliary controller or memory. Ignoring this auxiliary system leads to an apparent second-law violation, highlighting the importance of accounting for the thermodynamic costs of the controller's entropy reduction. The joint setup of controller and controlled system can only deliver positive net output work when the controller is at a lower temperature, thereby giving a heat engine that delivers net power by exchanging heat with two reservoirs at different temperatures. While previous theoretical analyses have hinted at this connection for specific systems [39–42], the results here are far broader, encompassing fully general mathematical proofs and intuitive explanations.

In addition to proving the main results using the theory of bipartite stochastic thermodynamics, I also illustrated them intuitively using an analogy to economics – providing a qualitative argument accessible without reference to stochastic thermodynamics. The concept of arbitrage lends itself well to understanding both classical and bipartite heat engines: the temperature of a heat bath can be thought of as the "exchange rate" between energy and entropy. By "trading" energy and entropy with different baths ("markets"), a heat engine can perform "arbitrage" to produce net output work ("profit"). Such a scheme requires moving energy and entropy from one bath to another, leading to the requirements for information flows and transduced power respectively quantified by the IFAR (7.14) and TPAR (7.15). The usefulness of this analogy should not be surprising; after all, thermodynamics is fundamentally the science of accounting for energy and entropy. Other analogies have likewise been drawn between stochastic thermodynamics and economics [116, 117]; exploring such analogies further, for example at a more quantitative level, could lead to deeper insights into both fields.

As shown in Sect. 7.3.2, the main results of this chapter [the IFAR (7.14) and TPAR (7.15)] can be reinterpreted as constraining changes in thermodynamic potentials of the environment encompassing the two thermal reservoirs. Since the reservoirs by assumption are at equilibrium, and do not directly support correlations or interactions with each other, it is straightforward to define their free energies (Helmholtz potentials) and free entropies (Massieu potentials) even when doing so for the system itself is not possible due to its nonequilibrium state and contact with multiple temperatures. These thermodynamic potentials are particularly useful, allowing for a qualitative understanding of the main results without the more involved theoretical machinery of stochastic thermodynamics; such an approach may prove useful for considering other nonequilibrium systems for which thermodynamic potentials cannot easily be defined.

Beyond elucidating design principles, the information-flow and transduced-power arbitrage relations (IFAR and TPAR) are also powerful tools for thermodynamic inference [118]. Using the IFAR (7.14), observing net output power ($\dot{W} < 0$) from a bipartite system immediately implies the existence of both a temperature difference and an information flow, whose sign is further implied if the ordering of the two temperatures is also known. Through TPAR (7.15), the existence and directionality of internally transduced power can likewise be inferred in autonomous systems at steady state. Chapter 8 showcases applications of the arbitrage relations for inference.

Finally, step back to consider macroscopic heat engines in simultaneous contact with two heat baths at different temperatures, for example thermoelectric devices. Since none of the theoretical results in Sect. 7.3 are built on assumptions about system size, the conclusions should still hold for macroscopic systems. This then raises the obvious question of where the information flow predicted by the IFAR can be found in, for example, a thermoelectric generator. I conjecture that the information flow is encoded in the statistics of electron positions and momenta [119]. A first step could be to consider small-scale systems which allow exact counting of electrons in a thermoelectric device. For larger thermoelectric devices, the electron positions and momenta are aggregated into correlations of voltage and current fluctuations across the two junctions in contact with different temperatures. Measurement of voltage and current fluctuations would be analogous to measuring the pressure in each quadrant of the two-dimensional ideal-gas Carnot engine in Sect. 7.2.2, which aggregates the statistics of the N gas molecules. The information flow in the engine could be extracted from pressure measurements in each quadrant with fine temporal resolution. It would be interesting to verify these predictions experimentally, namely measuring current and voltage correlations in a thermoelectric device, and thus quantifying the predicted information flow. Such a result would complement recent theoretical predictions of macroscopic information flows [119, 120], and serve to illustrate that core concepts from stochastic thermodynamics such as information flows have real relevance in macroscopic systems, far beyond the nanoscale regime in which they were originally formulated.

References

1. Piñol R, Zeler J, Brites CDS, Gu Y, Téllez P, Carneiro Neto AN, da Silva TE, Moreno-Loshuertos R, Fernandez-Silva P, Gellego AI, Marinez-Lostao L, Martínez A, Carlos LD, Millán A (2020) Real-Time intracellular temperature imaging using lanthanide bearing polymeric Micelles. Nano Lett 20:6466
2. Macherel D, Haraux F, Guillou H, Bourgeois O (2021) The conundrum of hot mitochondria. Biochimica et Biophysica Acta (BBA)-Bioenergetics 1862(2):148348
3. Wu T, Chen Z, Gong Z, Yan J, Guo J, Zhang Y, Li Y, Li B (2022) Intracellular thermal probing using aggregated fluorescent nanodiamonds. Adv Sci 9:2103354
4. Di X, Wang D, Su QP, Lio Y, Liao J, Maddahfar M, Zhou J, Jin D (2022) Spatiotemporally mapping temperature dynamics of lysosomes and mitochondria using cascade organelle-targeting upconversion nanoparticles. Proc Natl Acad Sci 119:e2207402119
5. Vinyard DJ, Ananyev GM, Dismukes GC (2013) Photosystem II: the reaction center of oxygenic photosynthesis. Ann Rev Biochem 82:577–606
6. Siddharth B, Parthiban S, Shanhui F (2018) Thermodynamic limits of energy harvesting from outgoing thermal radiation. Proc Natl Acad Sci 115(16):E3609–E3615
7. Emanuele P, Riccardo R, Massimiliano E (2021) Nonequilibrium thermodynamics of light-induced reactions. J Chem Phys 155(11):114101
8. Corra S, Bakić MT, Groppi J, Baroncini M, Silvi S, Penocchio E, Esposito M, Credi A (2022) Kinetic and energetic insights into the dissipative non-equilibrium operation of an autonomous light-powered supramolecular pump. Nature Nanotechnol 17(7):746–751
9. Mizuno D, Tardin C, Schmidt CF, MacKintosh FC (2007) Nonequilibrium mechanics of active cytoskeletal networks. Science 315:370
10. Gallet F, Arcizet D, Bohec P, Richert A (2009) Power spectrum of out-of-equilibrium forces in living cells: amplitude and frequency dependence. Soft Matter 5:2947
11. Pietzonka P, Fodor É, Lohrmann C, Cates ME, Seifert U (2019) Autonomous engines driven by active matter: energetics and design principles. Phys Rev X 9(4):041032
12. Lennart D, Stefano B, Ralf E (2019) Irreversibility in active matter systems: fluctuation theorem and mutual information. Phys Rev X 9(2):021009
13. Parry BR IV, Surovtsev MT Cabeen, O'Hern CS, Dufresne ER, Jacobs-Wagner C (2014) The bacterial cytoplasm has glass-like properties and is fluidized by metabolic activity. Cell 156(1–2):183–194
14. Guo M, Ehrlicher AJ, Jensen MH, Renz M, Moore JR, Goldman RD, Lippincott-Schwartz J, Mackintosh FC, Weitz DA (2014) Probing the stochastic, motor-driven properties of the cytoplasm using force spectrum microscopy. Cell 158(4):822–832
15. Fodor É, Guo M, Gov NS, Visco P, Weitz DA, Van Wijland F (2015) Activity-driven fluctuations in living cells. Europhys Lett 110(4):48005
16. Hartich D, Barato AC, Seifert U (2014) Stochastic thermodynamics of bipartite systems: transfer entropy inequalities and a Maxwell's demon interpretation. J Stat Mech Theory Exper 2014(2):P02016
17. Diana G, Esposito M (2014) Mutual entropy production in bipartite systems. J Stat Mech: Theory Exp 2014:P04010
18. Horowitz JM, Esposito M (2014) Thermodynamics with continuous information flow. Phys Rev X 4(3):031015
19. Ehrich J, Sivak D (2023) Energy and information flows in autonomous systems. Front Phys 11:1108357
20. Takayuki A, Keito T, Michio T, Daisuke M (2021) Noise-induced acceleration of single molecule kinesin-1. Phys Rev Lett 127(17):178101
21. Tripathi AK, Das T, Paneru G, Pak GK, Tlusty T (2022) Acceleration of enzymatic catalysis by active hydrodynamic fluctuations. Comm Phys 8:101
22. Grelier M, Sivak DA, Ehrich J (2024) Unlocking the potential of information flow: maximizing free-energy transduction in a model of an autonomous rotary molecular motor. Phys Rev E 109(3):034115

23. Ehrich J, Still S, Sivak DA (2023) Energetic cost of feedback control. Physical Review Research 5(2):023080
24. Rolf L (1961) Irreversibility and heat generation in the computing process. IBM J Res Dev 5(3):183–191
25. Crooks GE, Still S (2019) Marginal and conditional second laws of thermodynamics. EPL (Europhys Lett) 125(4):40005
26. Allahverdyan AE, Janzing D, Mahler G (2009) Thermodynamic efficiency of information and heat flow. J Stat Mech Theory Exper 2009(09):P09011
27. Horowitz JM, Gingrich TR (2020) Thermodynamic uncertainty relations constrain non-equilibrium fluctuations. Nature Phys 16(1):15–20
28. Still S, Sivak DA, Bell AJ, Crooks GE (2012) Thermodynamics of prediction. Phys Rev Lett 109:120604
29. Barato AC, Hartich D, Seifert U (2014) Efficiency of cellular information processing. New J Phys 16:103024
30. Hartich D, Barato AC, Seifert U (2016) Sensory capacity: an information theoretical measure of the performance of a sensor. Phys Rev E 93:022116
31. Large SJ, Ehrich J, Sivak DA (2021) Free-energy transduction within autonomous systems. Phys Rev E 103(2):022140
32. Lathouwers E, Sivak DA (2022) Internal energy and information flows mediate input and output power in bipartite molecular machines. Phys Rev E 105(2):024136
33. Amano S, Esposito M, Kreidt E, Leigh DA, Penocchio E, Roberts BMW (2022) Insights from an information thermodynamics analysis of a synthetic molecular motor. Nat Chem 14(5):530–537
34. Takaki R, Mugnai ML, Thirumalai D (2022) Information flow, gating, and energetics in dimeric molecular motors. Proc Natl Acad Sci 119:e2208083119
35. Leighton MP, Sivak DA (2023) Inferring subsystem efficiencies in bipartite molecular machines. Phys Rev Lett 130(17):178401
36. Paneru G, Dutta S, Pak HK (2022) Colossal power extraction from active cyclic Brownian information engines. J Phys Chem Lett 13:6912–6918
37. Malgaretti P, Stark H (2022) Szilard engines and information-based work extraction for active systems. Phys Rev Lett 129:228005
38. Saha TK, Ehrich J, Gavrilov M, Still S, Sivak DA, Bechhoefer J (2023) Information engine in a nonequilibrium bath. Phys Rev Lett 131:057101
39. Feynman RP, Leighton RB, Sands M (1965) The Feynman Lectures on Physics; vol. I. Am J Phys 33(9):750–752
40. Parrondo Juan MR, Pep E (1996) Criticism of Feynman's analysis of the ratchet as an engine. Am J Phys 64(9):1125–1130
41. Still S (2020) Thermodynamic cost and benefit of memory. Phys Rev Lett 124:050601
42. Still S, Daimer D (2022) Partially observable Szilard engines. New J Phys 24:073031
43. Leighton MP, Ehrich J, Sivak DA (2024) Information arbitrage in bipartite heat engines. Phys Rev X 14:041038
44. Samuelson PA, Nordhaus WD (2010) Economics, 19th edn. MaxGraw-Hill/Irwin, New York
45. Reich H (2012) Every force in nature (theory of everything, Part III). https://www.youtube.com/watch?v=3_RhISgoXUs
46. Ellerman DP (1984) Arbitrage theory: a mathematical introduction. SIAM Rev 26(2):241–261
47. Ellerman D (2000) Towards an arbitrage interpretation of optimization theory. World Bank
48. Léo T, Matteo M, Don Z (2021) Information thermodynamics of financial markets: the Glosten-Milgrom model. J Stat Mech: Theory Exp 3:033407
49. Ran H, Gabi S, Kornelius N (2018) Thermoelectric devices: a review of devices, architectures, and contact optimization. Adv Mater Technol 3(4):1700256
50. Gey M (2017) Exploiting Maxwell's demon: experiments on an information engine that can heat or cool. Bachelor Thesis, Universität Konstanz
51. Leo S (1929) Über die Entropieverminderung in einem thermodynamischen System bei Eingriffen intelligenter Wesen. Z Phys 53(11–12):840–856

52. Eric L, Sergio C (2015) Information: from Maxwell's demon to Landauer's eraser. Phys Today 68(9):30–35
53. Leff HS, Rex AF (eds) Maxwell's Demon 2. IOP Publishing
54. Parrondo JMR, Horowitz JM, Sagawa T (2015) Thermodynamics of information. Nat Phys 11:131
55. Callen HB (1998) Thermodynamics and an introduction to thermostatistics. American Association of Physics Teachers
56. Carnot S (1824) Reflections on the motive power of fire, and on machines fitted to develop that power. Paris: Bachelier 108:1824
57. Philipp S, Gernot S, Tobias B, Massimiliano E (2013) Thermodynamics of a physical model implementing a Maxwell demon. Phys Rev Lett 110(4):040601
58. Koski JV, Kutvonen A, Khaymovich IM, Ala-Nissila T, Pekola JP (2015) On-chip Maxwell's demon as an information-powered refrigerator. J Phys Rev Lett 115:260602
59. Ciliberto S (2020) Autonomous out-of-equilibrium Maxwell's demon for controlling the energy fluxes produced by thermal fluctuation. Phys Rev E 102:050103(R)
60. Freitas N, Esposito M (2021) Characterizing autonomous Maxwell demons. Phys Rev E 103:032118
61. Kay B, Keiji S, Udo S (2015) Thermodynamics of micro-and nano-systems driven by periodic temperature variations. Phys Rev X 5(3):031019
62. Toyabe S, Sagawa T, Ueda M, Muneyuki E, Sano M (2010) Experimental demonstration of information-to-energy conversion and validation of the generalized Jarzynski equality. Nat Phys 6(12):988
63. Camati PA, Peterson JPS, Batalhao TB, Micadei K, Souza AM, Sarthour RS, Oliveira IS, Serra RM (2016) Experimental rectification of entropy production by Maxwell's demon in a quantum system. Phys Rev Lett 117(24):240502
64. Cottet N, Jezouin S, Bretheau L, Campagne-Ibarcq P, Ficheux Q, Anders J, Auffèves A, Azouit R, Rouchon P, Huard B (2017) Observing a quantum Maxwell demon at work. Proc Natl Acad Sci 114(29):7561
65. Masuyama Y, Funo K, Murashita Y, Noguchi A, Kono S, Tabuchi Y, Yamazaki R, Ueda M, Nakamura Y (2018) Information-to-work conversion by Maxwell's demon in a superconducting circuit quantum electrodynamical system. Nat Commun 9(1):1
66. Koski JV, Maisi VF, Pekola JP, Averin DV (2014) Experimental realization of a Szilard engine with a single electron. Proc Natl Acad Sci 111(38):13786
67. Chida K, Desai S, Nishiguchi K, Fujiwara A (2017) Power generator driven by Maxwell's demon. Nat Commun 8(1):1
68. Tamir A, Saar R, Yael R (2018) Experimental realization of an information machine with tunable temporal correlations. Phys Rev Lett 121(18):180601
69. Paneru G, Lee DY, Tlusty T, Pak HK (2018) Lossless Brownian information engine. Phys Rev Lett 120(2):020601
70. Paneru G, Lee DY, Park J-M, Park JT, Noh JD, Pak HK (2018) Optimal tuning of a Brownian information engine operating in a nonequilibrium steady state. Phys Rev E 98(5):052119
71. Marco R-C, Felix R (2019) Large work extraction and the Landauer limit in a continuous Maxwell demon. Nat Phys 15(7):660–664
72. Paneru G, Dutta S, Sagawa T, Tlusty T, Pak HK (2020) Efficiency fluctuations and noise induced refrigerator-to-heater transition in information engines. Nature Commun 11(1):1–8
73. Saha TK, Lucero JNE, Ehrich J, Sivak DA, Bechhoefer J (2021) Maximizing power and velocity of an information engine. Proc Natl Acad Sci 118(20):e2023356118
74. Bauer M, Abreu D, Seifert U (2012) Efficiency of a Brownian information machine. J Phys A Math Theo 45:162001
75. Um J, Hinrichsen H, Kwon C, Park H (2015) Total cost of operating an information engine. New J Phys 17:085001
76. Schmitt RK, Parrondo JMR, Linke H, Johansson J (2015) Molecular motor efficiency is maximized in the presence of both power-stroke and rectification through feedback. New J Phys 17:065011

77. Bechhoefer J (2015) Hidden markov models for stochastic thermodynamics. New J Phys 17:075003
78. Lucero JNE, Ehrich J, Bechhoefer J, Sivak DA (2021) Maximal fluctuation exploitation in Gaussian information engines. Phys Rev E 104:044122
79. Cao FJ, Feito M (2009) Thermodynamics of feedback controlled systems. Phys Rev E 79:041118
80. Ponmurugan M (2010) Generalized detailed fluctuation theorem under nonequilibrium feedback control. Phys Rev E 82:031129
81. Horowitz JM, Vaikuntanathan S (2010) Nonequilibrium detailed fluctuation theorem for repeated discrete feedback. Phys Rev E 82:061120
82. Sagawa T, Ueda M (2012) Nonequilibrium thermodynamics of feedback control. Phys Rev E 85:021104
83. Sagawa T, Ueda M (2013) Role of mutual information in entropy production under information exchanges. New J Phys 15:125012
84. Ehrich J, Engel A (2017) Stochastic thermodynamics of interacting degrees of freedom: fluctuation theorems for detached path probabilities. Phys Rev E 96:042129
85. Roger F, Peter R (2007) Brownian gyrator: a minimal heat engine on the nanoscale. Phys Rev Lett 99(23):230602
86. Victor D, Anna M, Oleg V, Gleb O (2013) Two-temperature Langevin dynamics in a parabolic potential. Phys Rev E 87(6):062130
87. Andrea B, Andrea P, Luca S (2020) Engineered swift equilibration of a Brownian gyrator. Phys Rev E 102(3):030105
88. Miangolarra OM, Taghvaei A, Fu R, Chen Y, Georgiou TT (2021) Energy harvesting from anisotropic fluctuations. Phys Rev E 104(4):044101
89. Miangolarra OM, Taghvaei A, Georgiou TT (2024) Minimal entropy production in the presence of anisotropic fluctuations. IEEE Trans Autom Control
90. Sara C, Victor D, Gleb O, Lamberto R (2018) Asymmetry relations and effective temperatures for biased Brownian gyrators. Phys Rev E 98(4):042149
91. Fogedby HC, Imparato A (2017) A minimal model of an autonomous thermal motor. Europhys Lett 119(5):50007
92. Patrick P, Udo S (2018) Universal trade-off between power, efficiency, and constancy in steady-state heat engines. Phys Rev Lett 120(19):190602
93. Wenqi L, Yi-Hung L, Pik-Yin L, Yonggun J et al (2022) Stochastic currents and efficiency in an autonomous heat engine. Phys Rev E 106(2):L022106
94. Hsin C, Chi-Lun L, Pik-Yin L, Yung-Fu C (2021) Autonomous Brownian gyrators: a study on gyrating characteristics. Phys Rev E 103(2):022128
95. Youngkyoung B, Sangyun L, Juin K, Hawoong J (2021) Inertial effects on the Brownian gyrator. Phys Rev E 103(3):032148
96. Nascimento EDS, Morgado WAM (2021) Stationary properties of a non-markovian Brownian gyrator. J Stat Mech Theory Exper 2021(1):013301
97. Lee JS, Park J-M, Park H (2020) Brownian heat engine with active reservoirs. Phys Rev E 102(3):032116
98. Sergio C, Alberto I, Antoine N, Marius T (2013) Heat flux and entropy produced by thermal fluctuations. Phys Rev Lett 110(18):180601
99. Sergio C, Alberto I, Antoine N, Marius T (2013) Statistical properties of the energy exchanged between two heat baths coupled by thermal fluctuations. J Stat Mech: Theory Exp 12:P12014
100. Chiang K-H, Lee C-L, Lai P-Y, Chen Y-F (2017) Electrical autonomous Brownian gyrator. Phys Rev E 96(3):032123
101. Miangolarra OM, Taghvaei A, Chen Y, Georgiou TT (2022) Thermodynamic engine powered by anisotropic fluctuations. Physical Review Research 4(2):023218
102. Miangolarra OM, Taghvaei A, Georgiou TT (2023) A matching principle for power transfer in stochastic thermodynamics. IEEE Control Syst Lett 7:2107–2112
103. Aykut A, Jalpa S, Lennart D, Stefano B, Giuseppe P, Ralf E, Giovanni V (2017) Experimental realization of a minimal microscopic heat engine. Phys Rev E 96(5):052106

104. Abdoli I, Wittmann R, Brader JM, Sommer J-U, Löwen H, Sharma A (2022) Tunable Brownian magneto heat pump. Sci Rep 12(1):13405
105. Risken H (1996) Fokker-Planck equation. Springer
106. Maplesoft, a division of Waterloo Maple Inc.. Maple. https://hadoop.apache.org
107. Horowitz JM, Sandberg H (2014) Second-law-like inequalities with information and their interpretations. New J Phys 16(12):125007
108. Mingnan D, Fei L, Xiangjun X (2022) Unified theory of thermodynamics and stochastic thermodynamics for nonlinear Langevin systems driven by non-conservative forces. Phys Rev Res 4(4):043125
109. Alexander K, Harald S, Peter G (2004) Estimating mutual information. Phys Rev E 69(6):066138
110. Schaller G, Esposito M (2012) Stochastic thermodynamics for "Maxwell demon" feedbacks. EPL (Europhys Lett) 99:30003
111. Knott CG (1911) Life and Scientific Work of Peter Guthrie Tait. Cambridge University Press, London
112. Kutvonen A, Sagawa T, Ala-Nissila T (2016) Thermodynamics of information exchange between two coupled quantum dots. Phys Rev E 93:032147
113. Sánchez R, Samuelsson P, Potts PP (2019) Autonomous conversion of information to work in quantum dots. Phys Rev Res 1:033066
114. Annby-Andersson B, Samuelsson P, Maisi VF, Potts PP (2020) Maxwell's demon in a double quantum dot with continuous charge detection. Phys Rev B 101:165404
115. Tomohiro T, Van Tan V, Keiji S (2023) Universal bounds on the performance of information-thermodynamic engine. Phys Rev Res 5(4):043280
116. Matteo S (2016) Thermodynamics of inequalities: From precariousness to economic stratification. XXPhys A 441:40–50
117. Ducuara AF, Skrzypczyk P, Buscemi F, Sidajaya P, Scarani V (2023) Maxwell's demon walks into wall street: Stochastic thermodynamics meets expected utility theory. Phys Rev Lett 131(19):197103
118. Udo S (2019) From stochastic thermodynamics to thermodynamic inference. Ann Rev Condensed Matter Phys 10:171–192
119. Nahuel F, Massimiliano E (2023) Information flows in macroscopic Maxwell's demons. Phys Rev E 107(1):014136
120. Nahuel F, Massimiliano E (2022) Maxwell demon that can work at macroscopic scales. Phys Rev Lett 129(12):120602

Chapter 8
Information Arbitrage in Light-Harvesting Molecular Machines

Light-harvesting machines such as photosystem II and bacteriorhodopsin are in contact with both the thermal reservoir of the cellular interior, and blackbody radiation from the sun which can be treated as a thermal reservoir at a much higher temperature. I show that these molecular machines can be thought of as heat engines, leveraging differences in fluctuations to produce useful output work. The arbitrage relations derived in the previous chapter imply constraints on energy and information flows in bipartite heat engines; in this chapter I generalize those results, and highlight applications to thermodynamic inference by inferring information flows in photosystem II and bacteriorhodopsin.

8.1 Introduction

One example given to motivate the study of bipartite heat engines in the previous chapter was light-harvesting molecular machines like photosystem II [1] and bacteriorhodopsin [2]. These machines are driven out of equilibrium by high-energy solar photons, and the light-induced reactions can be treated as coupling to a heat bath at the temperature of the photon source [3–5]. In this chapter (Sects. 8.2 and 8.3) I will provide theoretical background for this counterintuitive claim, first giving a general argument based on the statistical mechanics of blackbody radiation, then examining in detail a simple example system.

The arbitrage relations introduced in the previous chapter have significant applications for thermodynamic inference, with light-harvesting molecular machines constituting a natural example. Since light-harvesting molecular machines cannot always be modelled as bipartite, Sect. 8.4 derives a general result called the entropy arbitrage relation (EAR) which holds for stochastic systems in contact with two thermal reservoirs without requiring the bipartite assumption. This allows for more general

M. Leighton, *Stochastic Thermodynamics of Multicomponent Molecular Machines*, Springer Theses, https://doi.org/10.1007/978-3-032-02204-2_8

application of arbitrage relations to light-harvesting molecular machines. Section 8.5 then applies the arbitrage relations to infer information flows in photosystem II and bacteriorhodopsin, verifying these predictions by comparing with published models.

8.2 Thermodynamics of Blackbody Radiation

The idea that light-energy can be thought of as heat from a radiation reservoir is highly counterintuitive, so I begin by sketching a proof for why energy exchanged with blackbody radiation or monochromatic radiation can be considered heat exchanged with a thermal reservoir at the temperature of the blackbody.

Suppose that a thermodynamic system can interact with photons of frequency ν that are at equilibrium with a blackbody at temperature T_{rad}, which also acts as a radiostat holding the photon number density constant. Following Ref. [4], the thermal radiation has number density (per unit volume) given by

$$n_\nu = \frac{f_\nu}{e^{\hbar\omega_\nu/k_B T_{\text{rad}}} - 1}, \tag{8.1a}$$

where $f_\nu = 2\omega_\nu^2/\pi c^3$ is the density of states.

The total energy density of photons at frequency ν is $U_\nu = n_\nu \hbar\omega_\nu$, which leads to a marginal energy change due to emission/absorption of a single photon of

$$u_\nu = \frac{\partial U_\nu}{\partial n_\nu} = \hbar\omega_\nu \tag{8.2}$$

as expected. Again following Ref. [4], the entropy density of photons at frequency ν is

$$S_\nu = [(f_\nu + n_\nu)\ln(f_\nu + n_\nu) - n_\nu \ln n_\nu - f_\nu \ln f_\nu]\,. \tag{8.3}$$

The marginal entropy change due to emission/absorption of a single photon is then

$$\begin{aligned} s_\nu &= \frac{\partial S_\nu}{\partial n_\nu} && (8.4a)\\ &= \ln\left(\frac{f_\nu + n_\nu}{n_\nu}\right) && (8.4b)\\ &= \frac{\hbar\omega_\nu}{k_B T_{\text{rad}}} && (8.4c)\\ &= \frac{u_\nu}{k_B T_{\text{rad}}}\,. && (8.4d) \end{aligned}$$

Equation (8.4d) is the crux of the argument, showing that the change in entropy of the radiation reservoir due to emission/absorption of a photon (i.e., the heat exchanged between the system and reservoir) is equal to the energy of the photon divided by the blackbody temperature. Thus energy the system exchanges with the radiation reservoir can be treated as heat exchanged with a thermal reservoir at temperature $T_{\rm rad}$.

Note that this argument holds also for systems interacting with monochromatic light, photons at a single frequency ν. This is because the energy and entropy exchanged with the radiation reservoir depend only on the number density of photons with frequency ν, not the rest of the intensity spectrum. Thus monochromatic radiation at a given frequency can be treated as a thermal reservoir at the temperature required to sustain the blackbody distribution corresponding to the observed intensity at frequency ν. In this way light-harvesting molecular machines, which interact with only specific wavelengths of light, can be treated as bipartite heat engines.

It is tempting to consider lasers, common laboratory sources of monochromatic light, as thermal reservoirs in this context; however, lasers are subject to added complications due to coherence as well as the unidirectional nature of the emitted light. A careful treatment of the thermodynamics of lasers is outside the scope of this thesis.

8.3 Illustrative Example

The thermodynamics of light-induced reactions are best illustrated with a simple example system, depicted in Fig. 8.1. Consider a molecule with two states, labeled $X = 0$ and $X = 1$, characterized by an energy difference $\Delta E = E_1 - E_0$. Suppose the molecule can transition between these states via two different mechanisms: a light-induced pathway mediated by interactions with photons of frequency ν, and a chemical pathway. To simplify the analysis, let the energy gap be equal to the energy of a photon with frequency ν, so that $\Delta E = \hbar\omega_\nu$.

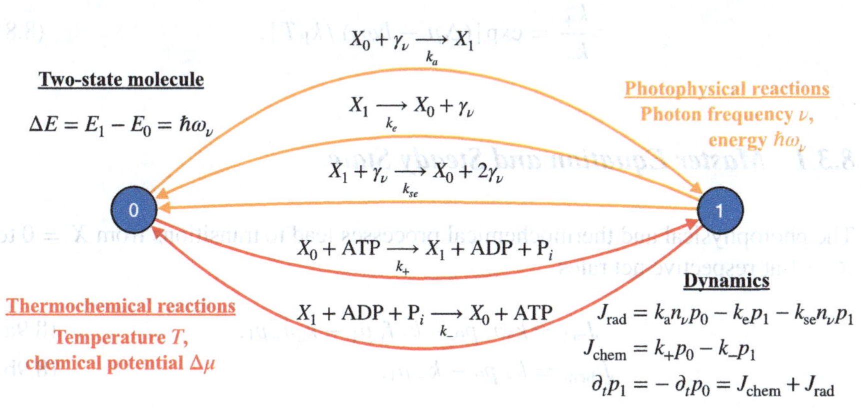

Fig. 8.1 Schematic illustrating the photophysical and thermochemical transitions in the simple example system

The molecule can undergo the following elementary light-induced transitions between states [6]:

$$X_0 + \gamma_\nu \longrightarrow X_1, \tag{8.5a}$$
$$X_1 \longrightarrow X_0 + \gamma_\nu, \tag{8.5b}$$
$$X_1 + \gamma_\nu \longrightarrow X_0 + 2\gamma_\nu. \tag{8.5c}$$

These photophysical processes correspond respectively to absorption, spontaneous emission, and stimulated emission, and occur at respective rates $k_\mathrm{a} n_\nu$, k_e, and $k_\mathrm{se} n_\nu$. Here n_ν is the number density of photons with frequency ν. Assuming the light is blackbody radiation, the rates of the different photophysical processes must obey [6]

$$k_\mathrm{se} = k_\mathrm{a}, \tag{8.6a}$$
$$k_\mathrm{e} = k_\mathrm{a} f_\nu. \tag{8.6b}$$

The molecule can also undergo the following chemical transitions, in contact with a thermal bath at temperature T:

$$X_0 + \mathrm{ATP} \longrightarrow X_1 + \mathrm{ADP} + \mathrm{P}_i, \tag{8.7a}$$
$$X_1 + \mathrm{ADP} + \mathrm{P}_i \longrightarrow X_0 + \mathrm{ATP}. \tag{8.7b}$$

Here the non-photo-induced transitions between molecular states are coupled to hydrolysis/catalysis of ATP, occurring at respective rates k_+ and k_-. The concentrations of ATP, ADP, and P_i are held fixed, with the ATP hydrolysis reaction having constant chemical potential $\Delta\mu$. To ensure thermodynamic consistency, these thermochemical transition rates are related by generalized detailed balance:

$$\frac{k_+}{k_-} = \exp\left[\left(\Delta\mu - \hbar\omega_\nu\right)/k_\mathrm{B}T\right]. \tag{8.8}$$

8.3.1 Master Equation and Steady State

The photophysical and thermochemical processes lead to transitions from $X = 0$ to $X = 1$ at respective net rates

$$J_\mathrm{rad} = k_\mathrm{a} n_\nu p_0 - k_\mathrm{a} f_\nu p_1 - k_\mathrm{a} n_\nu p_1, \tag{8.9a}$$
$$J_\mathrm{chem} = k_+ p_0 - k_- p_1. \tag{8.9b}$$

Thus the dynamics of the system are captured by the master equation

$$\partial_t p_1 = -\partial_t p_0 = J_{\text{rad}} + J_{\text{chem}}. \tag{8.10}$$

This master equation is easily solved, yielding steady-state probabilities

$$p_0 = \frac{k_- + k_{\text{a}} f_\nu + k_{\text{a}} n_\nu}{k_+ + k_- + k_{\text{a}}(2n_\nu + f_\nu)}, \tag{8.11a}$$

$$p_1 = \frac{k_+ + k_{\text{a}} n_\nu}{k_+ + k_- + k_{\text{a}}(2n_\nu + f_\nu)}. \tag{8.11b}$$

8.3.2 *Thermodynamics*

First Laws

I begin with the thermodynamics of the thermochemical transitions. The heat and chemical work rates (both defined to be positive when into the system) and change in system energy due to thermochemical transitions are

$$\dot{Q} = J_{\text{chem}}(\hbar\omega_\nu - \Delta_\mu), \tag{8.12a}$$

$$\dot{W}_{\text{chem}} = J_{\text{chem}}\Delta_\mu, \tag{8.12b}$$

$$\dot{E}_{\text{chem}} = J_{\text{chem}}\hbar\omega_\nu. \tag{8.12c}$$

Combined, these rates give a first law for thermochemical processes:

$$\dot{E}_{\text{chem}} = \dot{Q} + \dot{W}_{\text{chem}}. \tag{8.13}$$

Consider also the energetics of the photophysical transitions. The energy flow rate into the system from the photons is

$$\dot{U}_{\text{rad}} = J_{\text{rad}}\hbar\omega_\nu \tag{8.14a}$$

$$= \dot{E}_{\text{rad}}. \tag{8.14b}$$

The second line shows that the energy flow into the system from the blackbody reservoir is equal to the change $\dot{E}_{\text{rad}}$ in system energy due to the photophysical transitions.

Combining these statements yields a global first law describing energy conservation at the system level:

$$\mathrm{d}_t E = \dot{E}_{\text{chem}} + \dot{E}_{\text{rad}} \tag{8.15a}$$

$$= \dot{Q} + \dot{W}_{\text{chem}} + \dot{U}_{\text{rad}}. \tag{8.15b}$$

Global Second Law

Next consider the total entropy production rate due to the system dynamics, which is a sum of the time derivative of the system Shannon entropy, the rate of change in entropy of the thermal reservoir ($\dot{\mathcal{S}}_{\text{bath}}$), and the rate of change in entropy of the blackbody radiation ($\dot{\mathcal{S}}_{\text{rad}}$):

$$\dot{\Sigma} = \mathrm{d}_t S_{\text{sys}} + \dot{\mathcal{S}}_{\text{bath}} + \dot{\mathcal{S}}_{\text{rad}} \tag{8.16a}$$

$$= (J_{\text{rad}} + J_{\text{chem}}) \ln \frac{p_0}{p_1} - \frac{\dot{Q}}{k_{\text{B}}T} - J_{\text{rad}} s_\nu . \tag{8.16b}$$

The last term, $-J_{\text{rad}} s_\nu$, represents the rate of change of the blackbody radiation entropy due to absorption and emission of photons. Recalling the earlier result for s_ν for blackbody radiation at temperature T_{rad}, this term can be rewritten as

$$\dot{\mathcal{S}}_{\text{rad}} = -J_{\text{rad}} s_\nu \tag{8.17a}$$

$$= -J_{\text{rad}} \frac{\hbar\omega_\nu}{k_{\text{B}} T_{\text{rad}}} \tag{8.17b}$$

$$= -\frac{\dot{U}_{\text{rad}}}{k_{\text{B}} T_{\text{rad}}} . \tag{8.17c}$$

Again, this illustrates that the energy exchanged between the system and the thermal radiation can be treated as heat exchanged with a bath at temperature T_{rad}.

Second-Law Decomposition

The entropy production rate can be further decomposed into contributions due to each distinct transition mechanism. Defining the rates of system entropy change due to the photophysical ($\dot{S}_{\text{rad}}$) and thermochemical ($\dot{S}_{\text{chem}}$) transitions, it is then possible to define entropy production rates for each transition mechanism:

$$\dot{\Sigma}_{\text{rad}} = \dot{S}_{\text{rad}} - \frac{\dot{U}_{\text{rad}}}{k_{\text{B}} T_{\text{rad}}} , \tag{8.18a}$$

$$\dot{\Sigma}_{\text{chem}} = \dot{S}_{\text{chem}} - \frac{\dot{Q}}{k_{\text{B}} T} . \tag{8.18b}$$

Using these definitions, it is straightforward to show that these entropy production rates are both non-negative. For the photophysical transitions,

$$\dot{\Sigma}_{\text{rad}} = \dot{S}_{\text{rad}} - \frac{\dot{U}_{\text{rad}}}{k_{\text{B}}T_{\text{rad}}} \tag{8.19a}$$

$$= J_{\text{rad}} \ln \frac{p_0}{p_1} - J_{\text{rad}} \frac{u_\nu}{k_{\text{B}}T_{\text{rad}}} \tag{8.19b}$$

$$= J_{\text{rad}} \ln \frac{p_0}{p_1} - J_{\text{rad}} \ln \left(\frac{f_\nu + n_\nu}{n_\nu} \right) \tag{8.19c}$$

$$= J_{\text{rad}} \left[\ln \frac{p_0}{p_1} - \ln \left(\frac{k_a f_\nu + k_a n_\nu}{k_a n_\nu} \right) \right] \tag{8.19d}$$

$$= J_{\text{rad}} \ln \left(\frac{k_a n_\nu p_0}{k_a f_\nu p_1 + k_a n_\nu p_1} \right) \tag{8.19e}$$

$$= (k_a n_\nu p_0 - k_a f_\nu p_1 - k_a n_\nu p_1) \ln \left(\frac{k_a n_\nu p_0}{k_a f_\nu p_1 + k_a n_\nu p_1} \right). \tag{8.19f}$$

Here in Eq. (8.19c) I used Eq. (8.4d) to relate u_ν to n_ν and f_ν. The two multiplicative terms in the last line always have the same sign, so their product must be nonnegative. An analogous argument follows for $\dot{\Sigma}_{\text{chem}}$, so that the system satisfies two sub-second laws:

$$\dot{\Sigma}_{\text{rad}} = \dot{S}_{\text{rad}} - \frac{\dot{U}_{\text{rad}}}{k_{\text{B}}T_{\text{rad}}} \geq 0, \tag{8.20a}$$

$$\dot{\Sigma}_{\text{chem}} = \dot{S}_{\text{chem}} - \frac{\dot{Q}}{k_{\text{B}}T} \geq 0. \tag{8.20b}$$

Combining these inequalities with the transition-specific first laws and assuming steady state so that $\text{d}_t E = 0 = \text{d}_t S$, I derive an analog to the IFAR for this system,

$$-\dot{W}_{\text{chem}} \leq k_{\text{B}}(T_{\text{rad}} - T)\dot{S}_{\text{rad}}. \tag{8.21}$$

In the next section I will derive a far more general form of this result.

Figure 8.2 shows how this inequality constrains the chemical output work, as well as how it compares to the Carnot bound if this system is treated as a classical heat engine.

8.4 Generalized Entropy Arbitrage

Chapter 7 focused on bipartite heat engines composed of two distinct subsystems each in contact with a different thermal reservoir. In some cases however, it may not be possible to resolve distinct degrees of freedom corresponding to different subsystems, while still being able to ascribe the rates of different transitions to coupling with distinct reservoirs. An example would be a system described by a discrete set of states $\{z_i\}$, whose dynamics follow the master equation

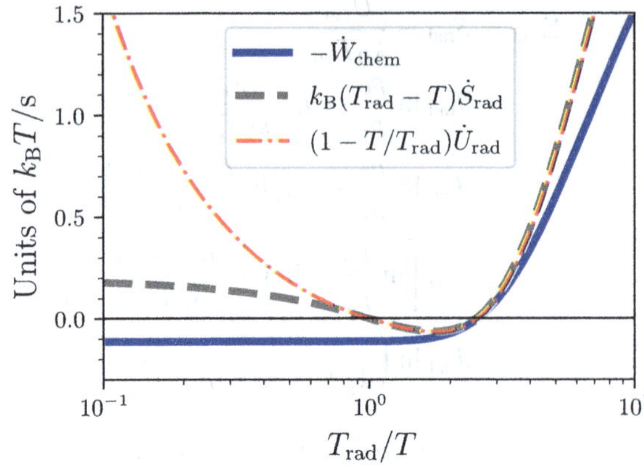

Fig. 8.2 Verification of the entropy arbitrage relation (EAR) and Carnot bound on output work in the simple photochemical system considered in Sect. 8.3. Parameters are $\beta\Delta\mu = 6$, $\beta\hbar\omega_v = 10$, $f_v = 1$, $k_a = 1000/\mathrm{s}$, and $k_- = 1/\mathrm{s}$

$$\frac{\partial}{\partial t} p_i = \sum_j \left[\left(k_{ij}^A + k_{ij}^B\right) p_j - \left(k_{ji}^A + k_{ji}^B\right) p_i \right] , \tag{8.22}$$

for transition rates k_{ij}^A and k_{ij}^B respectively coupled to distinct thermal reservoirs A and B with respective temperatures T_A and T_B. The system considered in Sect. 8.3 constitutes one example. Another example is bacteriorhodopsin, a light-harvesting molecular machine found in certain microorganisms which has recently been analyzed through the lens of stochastic thermodynamics [7], and is typically modelled using a set of discrete states, with only a single transition mediated by light [2].

8.4.1 Decomposing the Second Law Without the Bipartite Structure

Consider a stochastic system Z with a discrete set of states $\{z_i\}$, coupled to two distinct thermal reservoirs A and B, whose dynamics follows the master equation

$$\frac{\partial}{\partial t} p_i = \sum_j \left[\left(k_{ij}^A + k_{ij}^B\right) p_j - \left(k_{ji}^A + k_{ji}^B\right) p_i \right] . \tag{8.23}$$

Here k_{ij}^A and k_{ij}^B denote transition rates respectively coupled to the two distinct thermal reservoirs A and B at respective temperatures T_A and T_B.

The total entropy production rate, encompassing changes $\mathrm{d}_t S[Z]$ of the system Shannon entropy and heat flows $\dot{Q}_A$ and $\dot{Q}_B$ from the respective thermal reservoirs, is

$$\dot{\Sigma} = \underbrace{\sum_{i>j} \left[\left(k_{ij}^A + k_{ij}^B\right) p_j - \left(k_{ji}^A + k_{ji}^B\right) p_i \right] \ln \frac{p_j}{p_i}}_{\mathrm{d}_t S[Z]}$$
$$+ \underbrace{\sum_{i>j} \left[k_{ij}^A p_j - k_{ji}^A p_i \right] \ln \frac{k_{ij}^A}{k_{ji}^A}}_{-\dot{Q}_A/k_{\mathrm{B}} T_A} + \underbrace{\sum_{i>j} \left[k_{ij}^B p_j - k_{ji}^B p_i \right] \ln \frac{k_{ij}^B}{k_{ji}^B}}_{-\dot{Q}_B/k_{\mathrm{B}} T_B} . \tag{8.24}$$

This decomposition is similar to the analysis of Ref. [8], which similarly identifies entropy production contributions due to different reservoirs. By splitting the system entropy change into contributions due to transitions respectively coupled to the A and B reservoirs,

$$\mathrm{d}_t S[Z] = \underbrace{\sum_{i>j} \left[k_{ij}^A p_j - k_{ji}^A p_i \right] \ln \frac{p_j}{p_i}}_{\dot{S}_A} + \underbrace{\sum_{i>j} \left[k_{ij}^B p_j - k_{ji}^B p_i \right] \ln \frac{p_j}{p_i}}_{\dot{S}_B} , \tag{8.25}$$

the total entropy production rate can be decomposed into respective contributions due to the A and B reservoirs:

$$\dot{\Sigma} = \underbrace{\dot{S}_A - \frac{\dot{Q}_A}{k_{\mathrm{B}} T_A}}_{\dot{\Sigma}_A} + \underbrace{\dot{S}_B - \frac{\dot{Q}_B}{k_{\mathrm{B}} T_B}}_{\dot{\Sigma}_B} . \tag{8.26}$$

As in the bipartite case, here $\dot{\Sigma}_A$ and $\dot{\Sigma}_B$ are both nonnegative:

$$\dot{\Sigma}_A = \dot{S}_A - \frac{\dot{Q}_A}{k_{\mathrm{B}} T_A} \tag{8.27a}$$
$$= \sum_{i>j} \left(k_{ij}^A p_j - k_{ji}^A p_i \right) \ln \frac{p_j}{p_i} + \sum_{i>j} \left(k_{ij}^A p_j - k_{ji}^A p_i \right) \ln \frac{k_{ij}^A}{k_{ji}^A} \tag{8.27b}$$
$$= \sum_{i>j} \left(k_{ij}^A p_j - k_{ji}^A p_i \right) \ln \frac{k_{ij}^A p_j}{k_{ji}^A p_i} \tag{8.27c}$$
$$\geq 0. \tag{8.27d}$$

The inequality in the last line follows from noting that the difference and log-ratio must have the same sign, so that their product cannot be negative. The same logic holds for the entropy production rate due to transitions coupled to the B reservoir, so that

$$\dot{\Sigma}_B = \dot{S}_B - \frac{\dot{Q}_B}{k_{\mathrm{B}} T_B} \geq 0. \tag{8.28}$$

8.4.2 *Entropy Arbitrage Relation*

At steady state the system entropy is constant, so $0 = \mathrm{d}_t S = \dot{S}_A + \dot{S}_B$. Combining this with the reservoir-specific second laws and the steady-state first law $-\dot{W} = \dot{Q}_A + \dot{Q}_B$, I derive an IFAR-like result:

$$-\dot{W} \leq k_\mathrm{B}(T_A - T_B)\dot{S}_A. \tag{8.29}$$

We call this generalized result the *entropy arbitrage relation* (EAR), which holds for nonequilibrium systems in contact with two thermal reservoirs, without requiring the bipartite structure. When the system is bipartite, such that it can be decomposed into two subsystems each in contact with a unique reservoir (as detailed in Sect. 7.3), then the entropy rate $\dot{S}_A$ is equivalently the information flow from one subsystem to the other. Thus the IFAR emerges as the bipartite specialization of the EAR.

This inequality can be recast in terms of an information rate by considering the self-information $I[Z; Z] = S[Z]$ [9, Sect. 2.4], so that

$$-\dot{W} \leq k_\mathrm{B}(T_A - T_B)\dot{I}_A. \tag{8.30}$$

The information rate $\dot{I}_A$ is the rate at which the transitions coupled to the A reservoir change the self-information $I[Z; Z]$. We call this quantity an information rate (rather than a flow) because it cannot always be considered a flow of information from one part of the system to another.

8.5 Inferring Information Flows in Light-Harvesting Molecular Machines

I now turn to applying the arbitrage relations for inference, both qualitative and quantitative.

Using the IFAR (7.14), observing net output power ($\dot{W} < 0$) from a bipartite system immediately implies the existence of both a temperature difference and an information flow, whose sign is further implied if the ordering of the two temperatures is also known. Through TPAR (7.15), the existence and directionality of internally transduced power can likewise be inferred in autonomous systems at steady state. More quantitatively, each of the IFAR, TPAR, and EAR (8.30) can be rearranged to yield a bound on internal flows of energy or information, requiring only knowledge of (often experimentally tractable) input and output works and the two reservoir temperatures. Using the IFAR and TPAR respectively, the information and energy flows inside a bipartite heat engine are bounded by

$$\dot{I}_Y \geq \frac{-\dot{W}}{k_B(T_X - T_Y)}, \tag{8.31a}$$

$$\dot{E}_X \geq \frac{-\dot{W}_X/T_X - \dot{W}_Y/T_Y}{1/T_Y - 1/T_X}. \tag{8.31b}$$

I will illustrate this application by inferring the existence and magnitude of the information rate inside photosystem II, one of the molecular machines responsible for photosynthesis in plant cells. Photosystem II is in contact with the ambient cellular environment ($T_Y \approx 300\,\text{K}$) as well as hot solar radiation emitted from the surface of the sun at $\approx$5800 K [10]. Its dynamics include light-induced electronic transitions of the P680 complex (coupled to a high-temperature thermal reservoir at $T_X \leq 5800\,\text{K}$) and water-splitting chemical reactions of the oxygen-evolving complex (OEC) [1] coupled to the lower-temperature thermal reservoir at T_Y. While models of photosystem-II dynamics differ on whether the dynamics of P680 and OEC satisfy the bipartite assumption [11–13], they uniformly ascribe each transition to a particular reservoir. Estimating the mean output work from the free-energy change $\approx$237 kJ/mol [14] and net reaction rate $\approx$350/s [15], I infer a minimum information rate of $\approx$7 bit/reaction (or equivalently $\approx$2000 bit/s) inside photosystem II. In general this is an information rate due to the dynamics of light-induced transitions, however under the more restrictive bipartite assumptions of IFAR this is more specifically an information flow between the OEC and P680 (the rate is the same, only the interpretation differs).

While available experimental data for photosystem II is insufficient to directly quantify information flows due to photochemical dynamics, detailed stochastic models for the reaction dynamics have been constructed and fit to experimental data. A popular such model, Lazár and Jablonsky's Scheme 4 [13] (which can be described by a 24-state master equation of the form Eq. (8.22), detailed in Ref. [13, Appendix A.3], with the A and B reservoirs corresponding respectively to solar blackbody radiation, and to the ambient thermal bath of the cellular environment) incorporates both photophysical dynamics of the P680 complex and chemical dynamics of the OEC. This model is not bipartite, but uniquely identifies transitions as coupled to either photon absorption/emission or chemical reaction dynamics. Note that this model violates the assumption of thermodynamic consistency: several model transitions are irreversible. In this experimentally parameterized stochastic model for photosystem II dynamics, calculations yield an information rate of $\approx$9 bit/reaction, above but remarkably close to the model-agnostic lower bound.

In addition to photosystem II, arbitrage relations can be applied to other light-harvesting molecular machines like bacteriorhodopsin, which uses free energy from sunlight to pump protons across membranes in diverse species of archaea. The reaction dynamics and thermodynamics of bacteriorhodopsin are well understood [2, 16], and while the reaction cycle is not bipartite, models nonetheless uniquely couple transitions to either solar photons ($T_X \leq 5800\,\text{K}$ as in photosystem II) or the ambient cellular environment ($T_Y \approx 293\,\text{K}$). Using a typical output work rate of $\approx 6.1\,k_B T_{\text{cell}}$/cycle [7], the EAR predicts an information rate with magnitude

$\gtrsim$0.5 bit/cycle. Using the observed cycle time of ~11.5 cycles/s, this is an information rate of ~5 bit/s. Solving the master equation for the 7-state model used in Ref. [7] (detailed in Ref. [7, SI Sect. IIA]), I compute an information rate of 2 bit/cycle, in agreement with the model-agnostic prediction and comparable to but somewhat lower than that found in photosystem II.

8.6 Discussion

I demonstrated in this chapter that the arbitrage relations (IFAR, TPAR, and EAR) are powerful tools for thermodynamic inference [17]. In particular, using only information about temperatures and external work rates, these arbitrage relations can be used to infer the existence and magnitude of internal energy and information flows within molecular machines. I illustrated this potential for model-agnostic inference in Sect. 8.5 by estimating the magnitude of information flow in the molecular machine photosystem II, which Sects. 8.2 and 8.3 argued can be considered a bipartite heat engine coupled to both photons at the high temperature of the solar blackbody spectrum and chemical reactions at the much cooler ambient cellular temperature. The resulting prediction, a lower bound on the information flow of ≈7 bits per reaction cycle (≈2000 bit/s), is validated by computation of the information rate (≈9 bits per reaction cycle) in an experimentally parameterized stochastic model of the photosystem II reaction cycle [13]. These results are similarly applied to infer a significant information rate (≈5 bit/s) in another light-harvesting molecular machine, bacteriorhodopsin, which is likewise verified through computational modelling. The magnitude of information rates found in these light-harvesting molecular machines is striking when compared to other biological information rates like that underlying bacterial chemotaxis, estimated at 0.03 bit/s [18], and those found in biochemical signalling networks, on the order of bits per hour [19]. It would be interesting to explore systematic variation of this information rate across different classes of molecular machines.

Inference Tool: The IFAR (7.14) and TPAR (7.15) can be used to bound internal energy and information flows in molecular machines well-described as bipartite heat engines, like photosystem II.

This chapter also provides a pedagogical introduction to the stochastic thermodynamics of molecular machines with light-induced reactions. I hope this introduction will prove useful in providing intuition for future work in this area, as light-harvesting molecular machines are a relatively underexplored area of application for stochastic thermodynamics.

References

1. Vinyard DJ, Ananyev GM, Dismukes GC (2013) Photosystem II: the reaction center of oxygenic photosynthesis. Ann Rev Biochem 82:577–606
2. Lórenz-Fonfría VA, Kandori H (2009) Spectroscopic and kinetic evidence on how bacteriorhodopsin accomplishes vectorial proton transport under functional conditions. J Amer Chem Soc 131(16):5891–5901
3. Buddhiraju S, Santhanam P, Fan S (2018) Thermodynamic limits of energy harvesting from outgoing thermal radiation. Proc Natl Acad Sci 115(16):E3609–E3615
4. Penocchio E, Rao R, Esposito M (2021) Nonequilibrium thermodynamics of light-induced reactions. J Chem Phys 155(11):114101
5. Corra S, Bakić MT, Groppi J, Baroncini M, Silvi S, Penocchio E, Esposito M, Credi A (2022) Kinetic and energetic insights into the dissipative non-equilibrium operation of an autonomous light-powered supramolecular pump. Nature Nanotechnol 17(7):746–751
6. Einstein A (1916) Strahlung-emission und-absorption nach der quantentheorie. Verh d Deutsche Physik Ges 18:318–328
7. Piñero J, Solé R, Kolchinsky A (2024) Optimization of nonequilibrium free energy harvesting illustrated on bacteriorhodopsin. Phys Rev Res 6(1):013275
8. Esposito M (2012) Stochastic thermodynamics under coarse graining. Phys Rev E 85(4):041125
9. Cover TM, Thomas JA (2006) Elements of Information Theory, 2nd edn. Wiley-Interscience, Hoboken, NJ
10. Dorfman KE, Voronine DV, Mukamel S, Scully MO (2013) Photosynthetic reaction center as a quantum heat engine. Proc Natl Acad Sci 110(8):2746–2751
11. Lazár D (2003) Chlorophyll a fluorescence rise induced by high light illumination of dark-adapted plant tissue studied by means of a model of photosystem II and considering photosystem II heterogeneity. J Theor Biol 220(4):469–503
12. Jablonsky J, Lazar D (2008) Evidence for intermediate s-states as initial phase in the process of oxygen-evolving complex oxidation. Biophys J 94(7):2725–2736
13. Lazár D, Jablonskỳ J (2009) On the approaches applied in formulation of a kinetic model of photosystem II: different approaches lead to different simulations of the chlorophyll a fluorescence transients. J Theor Biol 257(2):260–269
14. Ye S, Ding C, Li C (2019) Artificial photosynthesis systems for catalytic water oxidation. In: Advances in inorganic chemistry, vol 74, pp 3–59. Elsevier
15. Forbush B, Kok B, McGloin MP (1971) Cooperation of charges in photosynthetic O2 evolution-II. Damping of flash yield oscillation, deactivation. Photochem Photobiol 14(3):307–321
16. Varo G, Lanyi JK (1991) Thermodynamics and energy coupling in the bacteriorhodopsin photocycle. Biochemistry 30(20):5016–5022
17. Seifert U (2019) From stochastic thermodynamics to thermodynamic inference. Ann Rev Condens Matter Phys 10:171–192
18. Mattingly HH, Kamino K, Machta BB, Emonet T (2021) Escherichia coli chemotaxis is information limited. Nat Phys 17(12):1426–1431
19. Cheong R, Rhee A, Wang CJ, Nemenman I, Levchenko A (2011) Information transduction capacity of noisy biochemical signaling networks. Science 334(6054):354–358

Chapter 9
Conclusion

This chapter concludes the thesis. Here I summarize the main results, and highlight their broader implications. Then, results from Chapters 6 and 8 are combined to constrain and compare the free energy transduction strategies used in distinct biological molecular machines. This leads to a selection of suggested future research directions to build on this thesis. I conclude with a final remark.

This thesis explored the dynamics and thermodynamics of muticomponent molecular machines through the lens of stochastic thermodynamics. I leveraged existing theoretical tools and developed new ones in order to uncover fundamental limits, infer experimentally inaccessible thermodynamic quantities, and ultimately learn design principles.

Chapter 2 reviewed the theoretical developments central to this thesis, including the thermodynamics of nonequilibrium steady states, bipartite stochastic thermodynamics, and thermodynamic inference. While traditional analyses consider free-energy flows into and out of a machine, the bipartite formalism provides a finer-grained view. Decomposing the machine into multiple coupled components resolves internal details—internal flows of energy and information that enable the machine's function.

Chapter 3 detailed the derivation of a key theoretical tool used throughout this thesis, the Jensen bounds. These bounds constrain both subsystem and total entropy production rates for multicomponent stochastic systems—key thermodynamic quantities that are typically experimentally inaccessible—in terms of measurable quantities like friction coefficients and mean coordinate rates of change. The Jensen bounds are used throughout this thesis to derive fundamental limits, and for thermodynamic inference.

M. Leighton, *Stochastic Thermodynamics of Multicomponent Molecular Machines*, Springer Theses, https://doi.org/10.1007/978-3-032-02204-2_9

Chapters 4 and 5 focused on collective motor-driven transport, a paradigmatic example of multicomponent assemblies of molecular machines. I first delved into a specific model (Chap. 4), which permitted analytic solution for arbitrary numbers N of motors. Examining the behaviour of this model reveals scaling laws describing how various performance metrics of interest depend on N, and performance trade-offs that guide the choice of N. While derived explicitly for a simple analytically tractable model, these general findings are robust to complications like rough energy landscapes and motor binding/unbinding dynamics. Chapter 5 then considered a far more general class of collective transport models, using the Jensen bound to derive performance bounds and Pareto frontiers that constrain all systems independent of details like interactions between motors, cargo, and their substrates.

I focused on thermodynamic inference in Chap. 6, using the subsystem Jensen bounds to derive both upper and lower bounds on subsystem efficiencies in bipartite molecular machines. I used these bounds to infer the efficiencies of F_o and F_1 in ATP synthase and of a kinesin motor pulling a diffusive cargo from limited experimental data. These efficiency metrics had never previously been measured or calculated in the machines' coupled, in vivo contexts.

Finally, Chaps. 7 and 8 considered molecular machines in contact with multiple sources of fluctuations. I showed in Chap. 7 that machines in such a context can act as heat engines, extracting energy from a difference in fluctuations to produce net output work, only by transducing information internally between their components. Conversely, an information engine can only output more work than consumed to power it if it has access to an (effective) temperature difference. These ideas can be understood through a powerful analogy to economics, and shed new light on commonly studied model systems. I then used inequalities derived in Chap. 7 for thermodynamic inference in Chap. 8, obtaining the first quantitative estimates of information flows in light-harvesting molecular machines.

The box below highlights what I believe to be some of the key takeaways from the work contained in this thesis, phrased as answers to the motivating questions from Chap. 1 in terms of fundamental limits, inference tools, and design principles. In the final sections of this chapter, I highlight what I believe to be important future directions to build on this work towards the goal of continuing to address these driving questions.

What fundamental limits do the laws of physics impose on the performance of nanoscale machines?

1. The Jensen bound quantifies a minimum entropic cost for motion against viscous friction. Most fundamentally, this entropic cost comes at the subsystem level.
2. Equations (5.11), (5.12), and (5.13) provide bounds on individual performance metrics for general collective-transport systems, while Eqs. (5.17) and (5.18) describe Pareto frontiers for systems with identical motors.
3. The net output work of a bipartite heat engine or information engine is limited by the product of the information flow and temperature difference.

What can we infer about the inner workings of molecular machines from limited experimental data?

1. The Jensen bound can be used to estimate subsystem and total entropy production rates from only measurements of average coordinate rates of change and friction coefficients.
2. Subsystem efficiency upper and lower bounds allow inference of subsystem efficiencies in their natural coupled context from limited experimental data.
3. The IFAR (7.14) and TPAR (7.15) can be used to bound internal energy and information flows in molecular machines well-described as bipartite heat engines, such as photosystem II.

What design principles can we learn from biology to facilitate future engineering of synthetic nanomachines?

1. The number N of motors in a collective transport system can be tuned to optimize different performance metrics of interest. $N \gg D_{\mathrm{m}}/D_{\mathrm{c}}$ maximizes velocity and precision at the cost of high power consumption and decreased efficiency, while $N \ll D_{\mathrm{m}}/D_{\mathrm{c}}$ maximizes efficiency but leads to slow and imprecise transport.
2. Performance of a collective transport system can be maximized by making the potential energy landscape as close as possible to quadratic.
3. A bipartite heat engine that extracts useful work from different sources of fluctuations must be designed to operate as an information engine.

9.1 Comparing Free-Energy Transduction Strategies in Distinct Machines

Recall Fig. 2.1 in Chap. 2, which illustrated several different possible operational modes for bipartite engines. I believe an intriguing question going forward, which I have only begun to address in this thesis, is to determine what operational modes are used by biological molecular machines in different contexts.

While internal energy and information flows and other thermodynamic quantities (including various measures of efficiency) can be computed from simulations of stochastic models for molecular machines, directly measuring these quantities in experiment is much more difficult, and frequently intractable for current techniques: experimental determination of such quantities currently requires thermodynamic inference [1]. Throughout this thesis, I have developed novel techniques for inferring thermodynamic quantities using the Jensen bounds and arbitrage relations.

Combining results from Chaps. 6 and 8 on thermodynamic inference allows us to sketch out a map of constraints on operational modes used by different biological molecular machines. Figure 9.1 shows inferred energy and information flows in ATP synthase, kinesin, and photosystem II. The Y and X subsystems correspond respectively to F_o and F_1 (ATP synthase), kinesin and its diffusive cargo (kinesin),

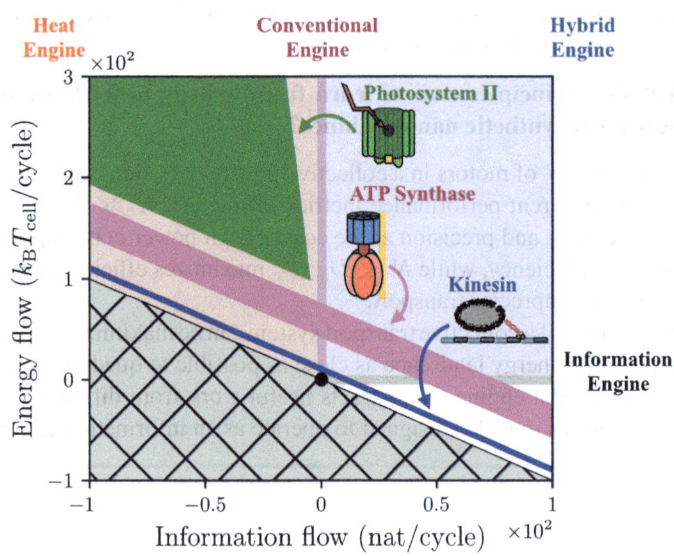

Fig. 9.1 Physical constraints on internal energy flows (relative to the thermal energy at cellular temperature T_{cell}) and information flows (in units of nats [2]) in molecular machines. Gray hatched region indicates negative free energy transduction $\dot{E}_Y + k_B T_{cell} \dot{I}_Y < 0$. Black circle: equilibrium ($\dot{E}_Y = 0 = \dot{I}_Y$). Light shaded regions show operational modes of heat, conventional, hybrid, and information engines (clockwise). Darker shaded regions show constraints on energy and information flows in the molecular machines photosystem II, ATP synthase, and kinesin

and the P680 and OEC subunits (photosystem II), assigned such that X is in contact with the cellular heat bath at T_{cell} and free energy flows from Y to X, hence $\dot{E}_Y + k_B T_{\text{cell}} \dot{I}_Y \geq 0$ (demarcated by the grey hatched "forbidden" region).

Mathematically, for ATP synthase and kinesin, Fig. 9.1 plots upper and lower bounds on the transduced free energy, the sum of the internal energy and information flows:

$$\beta\zeta_X\langle\dot{x}\rangle^2 - \beta\dot{W}_X \leq \beta\dot{E}_Y + \dot{I}_Y \leq \beta\dot{W}_Y - \beta\zeta_Y\langle\dot{y}\rangle^2. \tag{9.1}$$

For photosystem II, Fig. 9.1 plots inequalities constraining the internal energy flow in terms of the information flow, temperatures, and output work, obtained by rearranging Eq. (7.13) with $\dot{W}_Y = 0$:

$$-\beta_X\dot{W}_X - \dot{I}_Y \leq \beta_X\dot{E}_Y \leq -\frac{T_Y}{T_X}\dot{I}_Y. \tag{9.2}$$

These upper and lower bounds on $\dot{E}_Y$ intersect at $\dot{I}_Y$ equal to the IFAR lower bound. The parameters used are the same as the estimates discussed in Chaps. 6 and 8.

Figure 9.1 shows regions for heat, conventional, hybrid, and information engines, along with inferred constraints on flows inside select molecular machines. For photosystem II, the arbitrage relations constrain its internal thermodynamics to the heat-engine regime, while for ATP synthase and kinesin, constraints from the subsystem Jensen bounds leave open all four different operational modes as possibilities. Figure 9.1 also clearly illustrates how these machines differ in the magnitude of internal free-energy transduction (quantitatively, distance from the bounding line where $\dot{E}_Y + k_B T_{\text{cell}} \dot{I}_Y = 0$): photosystem II supports significantly more internal free-energy transduction per cycle than ATP synthase, which in turn out-transduces kinesin by a similar margin.

I hope to see future work add further constraints and other molecular machines to this figure. To accomplish this, future work should develop more and better tools for thermodynamic inference. While work to date, outlined in Sect. 2.8, has yielded constraints on the internal details of specific biological machines, these results are insufficient to determine the operational modes used in isothermal machines like kinesin and ATP synthase. Tighter bounds and alternative methods are needed to improve both the precision and accuracy of estimates. Likewise, future work should bring these inference tools to bear on other molecular machines beyond those that have been studied thus far. Exploring the internal details of different classes of molecular machines with improved inference tools will more systematically identify the free-energy transduction strategies used by different classes of evolved biological molecular machines (see Fig. 9.1) in different contexts, allowing identification of patterns and extraction of design principles for engineering synthetic molecular machines.

These efforts would also be aided by efforts to experimentally measure quantities that can be used for thermodynamic inference (Sect. 2.8), both making new measurements and improving the precision of existing ones. These include friction coefficients and rates of change for continuous degrees of freedom (e.g., angular coordinates in rotary motors), free-energetic driving forces, and variances in measured fluxes. Such measurements will enable inference-driven exploration of understudied and novel molecular machines and improve inference of the workhorse examples highlighted in this thesis.

Ultimately we should measure internal energy and information flows directly to test our theoretical predictions. Cutting-edge experiments are tantalizingly close to achieving the ability to track multiple degrees of freedom simultaneously with high spatiotemporal resolution. For example, dual-color MINFLUX shows significant promise [3], allowing simultaneous tracking of both heads of a kinesin motor. As this data becomes available, theorists should develop and benchmark different methods for computing thermodynamic quantities from noisy, limited data. In the longer term, as measurement techniques improve and data becomes increasingly plentiful and precise, computational resources should be built for rapid, efficient, and tractable analysis.

9.2 Leveraging Active Fluctuations

Chapters 7 and 8 explored the stochastic thermodynamics of molecular machines in contact with thermal reservoirs at different temperatures, which can give rise to different types of fluctuations acting on different parts of a multicomponent molecular machine. As mentioned in Chap. 7, biology features a wide range of active fluctuations [4–7]. While active fluctuations can in some cases be modeled as thermal fluctuations at a higher temperature, this is certainly not always the case, since they can have significant spatiotemporal correlations [4].

A general theory has yet to be developed for thermodynamic engines leveraging active fluctuations; nevertheless, explorations of specific systems have led to insights. Outside of molecular machines, recent work has shown that information can be used to leverage active fluctuations to produce useful work in active heat engines [8], active matter [9], and microorganisms swimming through turbulent flows [10]. Of particular interest, Ref. [8] derives a second-law-like inequality for active engines which, similar to the IFAR, bounds output work in terms of information-theoretic flows (albeit not directly related to the information flow considered in this thesis).

For molecular machines, active fluctuations as a resource remain relatively unexplored. Intriguing experiments and computation have shown that certain types of active noise can speed up kinesin operation when pulling a diffusive cargo against high load forces [11]. In future work, I hope to determine thermodynamic constraints on molecular machines with access to active fluctuations. I strongly suspect that an IFAR-like relationship will limit work extraction from active fluctuations in terms of information flow. Another important question will be to determine whether other molecular machines can similarly leverage active noise to improve their performance.

9.3 Other Future Directions

Many of the key results in this thesis rely on the Jensen bound (derived in Chap. 3), which only applies to stochastic systems with continuous dynamics. Discrete master-equation dynamics, the other main class of models for nanoscale stochastic processes, represent a formidable challenge for applying the Jensen bound. Since they do not feature bare friction/diffusion coefficients, or continuously varying coordinates whose mean rate of change can be measured, discrete dynamics are currently beyond the scope of the arguments presented here. Nevertheless, many discrete models of classical physical processes are really coarse-grained approximations of some underlying continuous dynamics, which would themselves obey the Jensen bound as derived in this article. Based on this, it seems likely that a discrete generalization of the Jensen bound is possible. The derivation of the Jensen bound also breaks for velocity-dependent forces (like those arising from magnetic fields), which are also known to break the TUR in some cases [12], or non-thermal sources of fluctuations like active noise [4, 5].

In this thesis, I primarily considered molecular machines that could either plausibly be modelled as bipartite, or multicomponent assemblies of many identical molecular machines. Many other molecular machines of significant biophysical interest, however, do not invite such a simple decomposition. For example, the bacterial flagellar motor is composed of >10 coupled components (rotor, stators, hook, etc...), while the ribosome is thought to comprise $\mathcal{O}(100)$ components [13]. As such, future theoretical work should seek to solidify and expand the framework of multipartite stochastic thermodynamics beyond the first and second laws described in Sect. 2.5, for example towards multipartite generalizations of subsystem efficiencies. While for bipartite systems the internal energy and information flows can be described as flowing from one subsystem to another, in a multipartite system such an interpretation is much less straightforward to make except in special cases where the subsystems are sparsely connected (such as the collective-transport system considered in Chap. 4). Going from a single system to a bipartite one introduces the possibility of internal energy and information flows; it is natural to wonder what new qualitative features arise when systems are decomposed further. I expect the topology of the graph describing interactions between components to play an important role.

The information flow discussed throughout this thesis is at this point fairly well understood mathematically, but still somewhat mysterious from the standpoint of physical intuition. For example, Chap. 8 inferred $\gtrsim 7$ bits information flow per reaction cycle in photosystem II. How should we interpret the magnitude of this information flow? The information flow can be interpreted in terms of measurement and feedback [14], but the connection to an autonomous molecular machine could be sharpened. Future work should develop intuition for the magnitude of information flows, perhaps drawing on connections between stochastic thermodynamics and computation [15].

Biological and synthetic molecular machines are characterized by their nanoscale size, stochastic degrees of freedom, and energy scales comparable to the thermal energy $k_B T$. As a result of these features information can be a relevant free-energy resource, showcased by the importance of information flows in internal free-energy transduction. An intriguing question is to what extent molecular machines which rely on information flows can be scaled up in size. Existing work has explored how information flows scale with system size in electronic systems [16, 17], and shown that macroscopic heat engines can be constructed using information flow [18]. Other work has likewise shown that biological organisms can leverage fluctuations in their environments at larger scales than the cellular interior. Bacteria [10] and birds [19] leverage nonequilibrium fluid flows for locomotion, and humans are known to do the same, for example in sailing, which requires collecting information about the prevailing winds. Future work should explore how features like internal energy and information flows might behave as molecular machines are scaled up in size towards the macroscopic regime.

9.4 Final Remark

Molecular machines feature centrally in free-energy transduction across biology, with multicomponent machines constituting many of the most prominent examples. Biological molecular machines have been tuned by billions of years of evolution, achieving remarkable performance that in many cases exceeds the best that human engineering has to offer. With the goal of learning from biology, this thesis has explored the stochastic thermodynamics of multicomponent molecular machines. This work derived fundamental limits, inferred hidden properties, and ultimately uncovered design principles for general nanomachines. While we have learned much, ultimately this work only begins to scratch the surface of understanding the intricate machinery hidden inside every living organism. I hope this thesis will serve as a launchpad for further study of free-energy transduction by biological molecular machines, and that the results within will prove useful in guiding efforts to engineer synthetic nanomachines.

References

1. Seifert U (2019) From stochastic thermodynamics to thermodynamic inference. Ann Rev Conden Matter Phys 10:171–192
2. Cover TM, Thomas JA (2006) Elements of information theory, 2nd edn. Wiley-Interscience, Hoboken, NJ
3. Scheiderer L, Wirth JO, Tarnawski M, Hell SW (2024) Dual-color minflux: kinesin-1 takes Chassé-inchworm steps. bioRxiv pp 2024–03
4. Mizuno D, Tardin C, Schmidt CF, MacKintosh FC (2007) Nonequilibrium mechanics of active cytoskeletal networks. Science 315:370
5. Gallet F, Arcizet D, Bohec P, Richert A (2009) Power spectrum of out-of-equilibrium forces in living cells: amplitude and frequency dependence. Soft Matter 5:2947

6. Pietzonka P, Fodor É, Lohrmann C, Cates ME, Seifert U (2019) Autonomous engines driven by active matter: energetics and design principles. Phys Rev X 9(4):041032
7. Dabelow L, Bo S, Eichhorn R (2019) Irreversibility in active matter systems: Fluctuation theorem and mutual information. Phys Rev X 9(2):021009
8. Datta A, Pietzonka P, Barato AC (2022) Second law for active heat engines. Phys Rev X 12(3):031034
9. VanSaders B, Vitelli V (2023) Informational active matter (2023). arXiv:2302.07402
10. Monthiller R, Loisy A, Koehl MAR, Favier B, Eloy C (2022) Surfing on turbulence: a strategy for planktonic navigation. Phys Rev Lett 129(6):064502
11. Ariga T, Tateishi K, Tomishige M, Mizuno D (2021) Noise-induced acceleration of single molecule kinesin-1. Phys Rev Lett 127(17):178101
12. Chun H-M, Fischer LP, Seifert U (2019) Effect of a magnetic field on the thermodynamic uncertainty relation. Phys Rev E 99(4):042128
13. Phillips R, Kondev J, Theriot J, Garcia H (2012) Physical biology of the cell. Garland Science
14. Ehrich J, Still S, Sivak DA (2023) Energetic cost of feedback control. Phys Rev Res 5(2):023080
15. Wolpert D, Korbel J, Lynn C, Tasnim F, Grochow J, Kardeş G, Aimone J, Balasubramanian V, de Giuli E, Doty D, et al (2023) Is stochastic thermodynamics the key to understanding the energy costs of computation? arXiv:2311.17166
16. Freitas N, Esposito M (2022) Maxwell demon that can work at macroscopic scales. Phys Rev Lett 129(12):120602
17. Freitas N, Esposito M (2023) Information flows in macroscopic Maxwell's demons. Phys Rev E 107(1):014136
18. Leighton MP, Ehrich J, Sivak DA (2024) Information arbitrage in bipartite heat engines. Phys Rev X 14:041038
19. Reddy G, Celani A, Sejnowski TJ, Vergassola M (2016) Learning to soar in turbulent environments. Proc Natl Acad Sci 113(33):E4877–E4884

Appendix A
Supplementary Material for Chap. 5

A.1 Linear Systems Saturate the Jensen Bound

Here I prove that a collective-transport system with only linear forces saturates the Jensen bound on entropy production (5.7c). Consider a linear system composed of $N+1$ subsystems with positions denoted $\{x_1, ..., x_{N+1}\}$, for the first N subsystems the motors, and the last the cargo, so $x_{N+1} \equiv x_c$ and $D_{N+1} \equiv D_c$. The system has constant force vector $\boldsymbol{f}$ and potential

$$V(\boldsymbol{x}) = V_0 + \sum_{i=1}^{N} \sum_{j=i+1}^{N+1} \kappa_{ij}(x_i - x_j)^2. \tag{A.1}$$

I neglect linear terms in the potential since they can be incorporated into the constant forces, and do not allow terms of the form $k_i x_i^2$ that depend on the absolute position of one subsystem, since they preclude the existence of a nonequilibrium steady state. The cargo may in general be subject to a non-zero external force, $f_{N+1} \equiv f_{\text{ext}}$.

The dynamics of this system are most simply written in Langevin form as

$$\dot{\boldsymbol{x}} = \beta \boldsymbol{D}\left[\boldsymbol{f} - \boldsymbol{A}\boldsymbol{x}\right] + \boldsymbol{\xi}(t). \tag{A.2}$$

Here $\boldsymbol{D}$ is the diffusivity matrix which, under the assumption of multipartite dynamics, is diagonal with entries $D_{ij} = D_i \delta_{ij}$, for Kroneker delta-function δ_{ij}. The matrix $\boldsymbol{A}$ satisfies $A_{ij} = \partial_{x_i} \partial_{x_j} V(\boldsymbol{x})$, and the vector-valued random noise $\boldsymbol{\xi}(t)$ has zero mean and covariance matrix

$$\left\langle \boldsymbol{\xi}(t)\, \boldsymbol{\xi}^\top(t') \right\rangle = 2\boldsymbol{D}\delta(t - t'). \tag{A.3}$$

The solution is a multivariate Gaussian distribution with mean vector $\boldsymbol{m}$ and covariance matrix $\boldsymbol{C} = \langle \boldsymbol{x}\boldsymbol{x}^\top - \boldsymbol{m}\boldsymbol{m}^\top \rangle$ satisfying the differential equations [1, Sect. 3.2]

M. Leighton, *Stochastic Thermodynamics of Multicomponent Molecular Machines*, Springer Theses, https://doi.org/10.1007/978-3-032-02204-2

$$\dot{\boldsymbol{m}} = \beta \boldsymbol{D}\left[\boldsymbol{f} - \boldsymbol{A}\boldsymbol{m}\right], \tag{A.4a}$$

$$\dot{\boldsymbol{C}} = -\beta \boldsymbol{D}\left[\boldsymbol{A}\boldsymbol{C} + \boldsymbol{A}^{\top}\boldsymbol{C}\right] + 2\boldsymbol{D}. \tag{A.4b}$$

By definition $\boldsymbol{A}$ and $\boldsymbol{C}$ are symmetric, so $\boldsymbol{A} = \boldsymbol{A}^{\top}$, $\boldsymbol{C} = \boldsymbol{C}^{\top}$, and $\boldsymbol{C}^{-1} = \left(\boldsymbol{C}^{-1}\right)^{\top}$.

The entropy production rate for the ith subsystem is [2]:

$$\dot{\Sigma}_i = \frac{1}{D_i}\left\langle\left(\frac{J_i(\boldsymbol{x},t)}{P(\boldsymbol{x},t)}\right)^2\right\rangle \tag{A.5a}$$

$$= \frac{1}{D_i}\left\langle\frac{1}{P(\boldsymbol{x},t)^2}\left(\beta D_i f_i P(\boldsymbol{x},t) - \beta D_i (\boldsymbol{A}\boldsymbol{x})_i P(\boldsymbol{x},t) - D_i \frac{\partial}{\partial x_i} P(\boldsymbol{x},t)\right)^2\right\rangle \tag{A.5b}$$

$$= \frac{1}{D_i}\left\langle\left(\beta D_i f_i - \beta D_i (\boldsymbol{A}\boldsymbol{x})_i - D_i \frac{\partial}{\partial x_i}\ln P(\boldsymbol{x},t)\right)^2\right\rangle \tag{A.5c}$$

$$= \frac{1}{D_i}\Bigg\langle\underbrace{(\beta D_i f_i)^2 - 2(\beta D_i)^2 f_i (\boldsymbol{A}\boldsymbol{x})_i + (\beta D_i (\boldsymbol{A}\boldsymbol{x})_i)^2}_{1} \tag{A.5d}$$

$$\underbrace{-2\beta(D_i)^2\left[f_i - (\boldsymbol{A}\boldsymbol{x})_i\right]\frac{\partial}{\partial x_i}\ln P(\boldsymbol{x},t)}_{2}\underbrace{+D_i^2\left[\frac{\partial}{\partial x_i}\ln P(\boldsymbol{x},t)\right]^2}_{3}\Bigg\rangle.$$

For clarity, I separately evaluate the three terms in this lengthy expression.

The first term is

$$\frac{1}{D_i}\left\langle(\beta D_i f_i)^2 - 2(\beta D_i)^2 f_i (\boldsymbol{A}\boldsymbol{x})_i + (\beta D_i (\boldsymbol{A}\boldsymbol{x})_i)^2\right\rangle$$

$$= \frac{(\beta D_i)^2}{D_i}\left[f_i^2 - 2 f_i (\boldsymbol{A}\boldsymbol{m})_i + \left\langle(\boldsymbol{A}\boldsymbol{x})_i^2\right\rangle\right] \tag{A.6a}$$

$$= \frac{(\beta D_i)^2}{D_i}\left[f_i^2 - 2 f_i (\boldsymbol{A}\boldsymbol{m})_i + \left\langle\boldsymbol{A}\boldsymbol{x}\boldsymbol{x}^{\top}\boldsymbol{A}\right\rangle_{ii}\right] \tag{A.6b}$$

$$= \frac{(\beta D_i)^2}{D_i}\left[f_i^2 - 2 f_i (\boldsymbol{A}\boldsymbol{m})_i + \left(\boldsymbol{A}\boldsymbol{m}\boldsymbol{m}^{\top}\boldsymbol{A}\right)_{ii} + (\boldsymbol{A}\boldsymbol{C}\boldsymbol{A})_{ii}\right] \tag{A.6c}$$

$$= \frac{1}{D_i}\left[(\beta D_i)^2\left(f_i - (\boldsymbol{A}\boldsymbol{m})_i\right)^2\right] + \beta^2 D_i (\boldsymbol{A}\boldsymbol{C}\boldsymbol{A})_{ii} \tag{A.6d}$$

$$= \frac{(\dot{\boldsymbol{m}}_i)^2}{D_i} + \beta^2 D_i (\boldsymbol{A}\boldsymbol{C}\boldsymbol{A})_{ii} \tag{A.6e}$$

$$= \frac{\langle v\rangle^2}{D_i} + \beta D_i\,(\boldsymbol{A})_{ii}\,. \tag{A.6f}$$

In the last line I took the steady-state limit so that $\dot{\boldsymbol{m}}_i = \langle v \rangle$. I further assumed that in the steady-state limit each term in the covariance matrix is linear in t:

$$\boldsymbol{C} = \boldsymbol{u}t + \boldsymbol{v}, \tag{A.7}$$

where both $\boldsymbol{u}$ and $\boldsymbol{v}$ must be symmetric. This linearity in t is necessary to obtain a constant effective diffusivity in the steady-state limit. The differential equation (A.4b) for the covariance then simplifies to

$$\boldsymbol{u} = -2\beta \boldsymbol{D}\boldsymbol{A}\boldsymbol{u}t - 2\beta \boldsymbol{D}\boldsymbol{A}\boldsymbol{v} + 2\boldsymbol{D}. \tag{A.8a}$$

Since the left-hand side is independent of t, the right-hand side must be as well. For this to be true for general $\boldsymbol{D}$, we must have $\boldsymbol{A}\boldsymbol{u} = 0$. I then evaluate the rightmost term in (A.6e):

$$\begin{aligned}
\boldsymbol{A}\boldsymbol{C}\boldsymbol{A} &= \beta^{-1}\boldsymbol{A} - \frac{1}{2}\beta^{-1}\boldsymbol{D}^{-1}\dot{\boldsymbol{C}}\boldsymbol{A} && \text{(A.9a)}\\
&= \beta^{-1}\boldsymbol{A} - \frac{1}{2}\beta^{-1}\boldsymbol{D}^{-1}\boldsymbol{u}\boldsymbol{A} && \text{(A.9b)}\\
&= \beta^{-1}\boldsymbol{A} - \frac{1}{2}\beta^{-1}\boldsymbol{D}^{-1}(\boldsymbol{A}\boldsymbol{u})^\top && \text{(A.9c)}\\
&= \beta^{-1}\boldsymbol{A}. && \text{(A.9d)}
\end{aligned}$$

The second term in (A.5d) is

$$\begin{aligned}
&\frac{1}{D_i}\left\langle -2\beta(D_i)^2\left[f_i - (\boldsymbol{A}\boldsymbol{x})_i\right]\frac{\partial}{\partial x_i}\ln P(\boldsymbol{x},t)\right\rangle\\
&= -2\frac{\beta D_i^2}{D_i}\left\langle \left[f_i - (\boldsymbol{A}\boldsymbol{x})_i\right]\frac{\partial}{\partial x_i}\left(-\frac{1}{2}(\boldsymbol{x}-\boldsymbol{m})^\top\boldsymbol{C}^{-1}(\boldsymbol{x}-\boldsymbol{m})\right)\right\rangle && \text{(A.10a)}\\
&= -2\frac{\beta D_i^2}{D_i}\left\langle \left[f_i - (\boldsymbol{A}\boldsymbol{x})_i\right]\left(-\frac{1}{2}\left[(\boldsymbol{x}-\boldsymbol{m})^\top\boldsymbol{C}^{-1}\right]_i - \frac{1}{2}\left[\boldsymbol{C}^{-1}(\boldsymbol{x}-\boldsymbol{m})\right]_i\right)\right\rangle && \text{(A.10b)}\\
&= 2\beta D_i\left\langle \left[f_i - (\boldsymbol{A}\boldsymbol{x})_i\right]\left(\boldsymbol{C}^{-1}(\boldsymbol{x}-\boldsymbol{m})\right)_i\right\rangle && \text{(A.10c)}\\
&= 2\beta D_i f_i\left\langle\left(\boldsymbol{C}^{-1}(\boldsymbol{x}-\boldsymbol{m})\right)_i\right\rangle - 2\beta D_i\left\langle \boldsymbol{A}\boldsymbol{x}\left(\boldsymbol{C}^{-1}(\boldsymbol{x}-\boldsymbol{m})\right)_i\right\rangle && \text{(A.10d)}\\
&= -2\beta D_i\left\langle\left(\boldsymbol{A}\boldsymbol{x}(\boldsymbol{x}-\boldsymbol{m})^\top\boldsymbol{C}^{-1}\right)_{ii}\right\rangle && \text{(A.10e)}\\
&= -2\beta D_i\left\langle\left(\boldsymbol{A}\boldsymbol{C}\boldsymbol{C}^{-1}\right)_{ii}\right\rangle && \text{(A.10f)}\\
&= -2\beta D_i\,(\boldsymbol{A})_{ii}\,. && \text{(A.10g)}
\end{aligned}$$

Finally, the third term in (A.5d) is

$$
\begin{aligned}
\frac{1}{D_i}\left\langle D_i^2\left[\frac{\partial}{\partial x_i}\ln P(\boldsymbol{x},t)\right]^2\right\rangle &= D_i\left\langle\left[-\frac{1}{2}\frac{\partial}{\partial x_i}\left((\boldsymbol{x}-\boldsymbol{m})^\top \boldsymbol{C}^{-1}(\boldsymbol{x}-\boldsymbol{m})\right)\right]^2\right\rangle && \text{(A.11a)}\\
&= D_i\left\langle\left(\boldsymbol{C}^{-1}(\boldsymbol{x}-\boldsymbol{m})\right)_i^2\right\rangle && \text{(A.11b)}\\
&= D_i\left\langle\left(\boldsymbol{C}^{-1}(\boldsymbol{x}-\boldsymbol{m})(\boldsymbol{x}-\boldsymbol{m})^\top \boldsymbol{C}^{-1}\right)_{ii}\right\rangle && \text{(A.11c)}\\
&= D_i\left(\boldsymbol{C}^{-1}\left\langle(\boldsymbol{x}-\boldsymbol{m})(\boldsymbol{x}-\boldsymbol{m})^\top\right\rangle \boldsymbol{C}^{-1}\right)_{ii} && \text{(A.11d)}\\
&= D_i\left(\boldsymbol{C}^{-1}\boldsymbol{C}\boldsymbol{C}^{-1}\right)_{ii} && \text{(A.11e)}\\
&= D_i\left(\boldsymbol{C}^{-1}\right)_{ii} && \text{(A.11f)}\\
&= \beta D_i(\boldsymbol{A})_{ii}. && \text{(A.11g)}
\end{aligned}
$$

To derive the last line I used

$$
\begin{aligned}
\boldsymbol{C}^{-1} &= \boldsymbol{A}\boldsymbol{A}^{-1}\boldsymbol{C}^{-1}\boldsymbol{A}^{-1}\boldsymbol{A} && \text{(A.12a)}\\
&= \boldsymbol{A}\left(\boldsymbol{A}\boldsymbol{C}\boldsymbol{A}\right)^{-1}\boldsymbol{A} && \text{(A.12b)}\\
&= \boldsymbol{A}\left(\beta^{-1}\boldsymbol{A}\right)^{-1}\boldsymbol{A} && \text{(A.12c)}\\
&= \beta\boldsymbol{A}\boldsymbol{A}^{-1}\boldsymbol{A} && \text{(A.12d)}\\
&= \beta\boldsymbol{A}. && \text{(A.12e)}
\end{aligned}
$$

Summing the three terms (A.6f), (A.10g), and (A.11g), the entropy production rate of the ith subsystem is

$$
\begin{aligned}
\dot{\Sigma}_i &= \frac{1}{D_i}\langle v\rangle^2 + \beta D_i\,(\boldsymbol{A})_{ii} - 2\beta D_i\,(\boldsymbol{A})_{ii} + \beta D_i\,(\boldsymbol{A})_{ii} && \text{(A.13a)}\\
&= \frac{1}{D_i}\langle v\rangle^2. && \text{(A.13b)}
\end{aligned}
$$

Thus the total rate of entropy production is

$$
\begin{aligned}
\dot{\Sigma} &= \sum_{i=1}^{N+1}\dot{\Sigma}_i\\
&= \left(\sum_{i=1}^{N+1}\frac{1}{D_i}\right)\langle v\rangle^2\\
&= \left(\frac{1}{D_\mathrm{c}}+\sum_{i=1}^{N}\frac{1}{D_i}\right)\langle v\rangle^2\\
&= \frac{\langle v\rangle^2}{D_\mathrm{bare}},
\end{aligned}
\qquad \text{(A.14)}
$$

exactly saturating the Jensen bound (5.7c).

A.2 Barrier Heights for Kinesin Motors

Here I use experimental data to estimate the heights of energy barriers separating metastable states for kinesin motors. Recall from (5.23) that the ith motor has a periodic potential $V_i(x_i) = \frac{1}{2}E^\ddagger \cos(2\pi x_i/\ell)$ with barrier height $E^\ddagger$, period ℓ, and maximum conservative force $f_{\max} = E^\ddagger/(2\ell)$. The Kramers rate [3] for an uncoupled motor hopping forward from one energy minimum to the next is

$$k_+ = \frac{\beta D_\mathrm{m}}{2\pi}\sqrt{\left|\frac{\partial^2 V_i}{\partial x_i^2}\right|_{x_i=a} \cdot \left|\frac{\partial^2 V_i}{\partial x_i^2}\right|_{x_i=b}}\, e^{-\beta E_b^+} \tag{A.15a}$$

$$\approx \frac{\pi\beta D_\mathrm{m} E^\ddagger}{\ell^2}\, e^{-\beta E_b^+}, \tag{A.15b}$$

where a is the position of the bottom of the current potential minimum and b is the position of the peak of the energy barrier to the right. The effective barrier height is $E_b^+ = E^\ddagger - f_\mathrm{chem}\ell/2$. Note that the cosine potential has second derivative of approximate magnitude $2\pi^2 E^\ddagger/\ell^2$ at both minima and peaks (points a and b) for $f_{\max}$ larger than f_chem.

Likewise, the rate for the motor hopping backward to the previous minimum is

$$k_- = \frac{\pi\beta D_\mathrm{m} E^\ddagger}{\ell^2}\, e^{-\beta E_b^-}, \tag{A.16}$$

where this time the effective barrier height is $E_b^- = E^\ddagger + f_\mathrm{chem}\ell/2$.

Analysis of experimental data [4] yields step rates for kinesin of $k_+ = 133.0/\mathrm{s}$ and $k_- = 0.2/\mathrm{s}$ [5], and step size $\ell = 8.2$ nm. Combining these with previous estimates of the motor diffusivity $D_\mathrm{m} \approx \mathcal{O}(10^{-3})\ \mu\mathrm{m}^2/\mathrm{s}$ [6, 7], solving the two equations (A.15b) and (A.16) for the two remaining parameters yields estimates $f_\mathrm{chem}\ell \approx 7\, k_\mathrm{B}T$ and $E^\ddagger = 2 f_{\max}\ell \approx 6\, k_\mathrm{B}T$. Accordingly, $f_{\max}/f_\mathrm{chem} \approx 0.4$ sets the scale for numerical investigations.

References

1. Risken H (1996) Fokker-Planck equation. Springer
2. Horowitz JM (2015) Multipartite information flow for multiple Maxwell demons. J Stat Mech: Theory Exper 2015(3):P03006
3. Hendrik Anthony Kramers (1940) Brownian motion in a field of force and the diffusion model of chemical reactions. Physica 7(4):284–304
4. Carter NJ, Cross RA (2005) Mechanics of the kinesin step. Nature 435(7040):308–312
5. Vu HT, Chakrabarti S, Hinczewski M, Thirumalai D (2016) Discrete step sizes of molecular motors lead to bimodal non-Gaussian velocity distributions under force. Phys Rev Lett 117(7):078101
6. Brown AI, Sivak DA (2019) Pulling cargo increases the precision of molecular motor progress. EPL (Europhys Lett) 126(4):40004
7. Leighton MP, Sivak DA (2022) Performance scaling and trade-offs for collective motor-driven transport. New J Phys 24(1):013009

9783032022035